The World Record Paper Airplane

and International Award Winning Designs:
The Best of John M. Collins and More
Paper Airplane Book

John M. Collins
The Paper Airplane Guy

The World Record Paper Airplane

and International Award Winning Designs:
The Best of John M. Collins and More
Paper Airplane Book

John M. Collins
The Paper Airplane Guy

To those who find joy in small things every day,
And to those who will learn to do so;
There is no other day like today.
Now is all we have. Live here.

Be kind when no one is watching.

THE PLANES

8 — Phoenix

14 — Suzanne

21 — The Plane

27 — Stealth

32 — Phat Glider

37 — Tumbling Wing

41 — Maple Seed

44 — RB Dart

48 — The Tube

51 — The Swan

56 — Bat Plane

62 — Ultra Glide

67 — Max Lock

Star Fighter — 73

Twin Jet — 78

Bonus Plane — 85

Interlock Dart — 88

Interlock Biplane — 93

Boomerang I — 96

Boomerang II — 101

FFF-1 — 105

Pelican — 109

Starship Shuttle — 113

Ring Thing — 118

Super Canard — 122

Seagull — 128

Flight Theory 133-148   Basic Folds 149-150   Folding Symbols 7 and inside back cover

Introduction

This superb collection of planes has flown on four continents for my educational paper airplane shows. Many of the designs are used in the aerobatic competition at Red Bull Paper Wings International Finals in Salzburg, Austria. If you'd like to compete there someday, this book is a great place to start.

Of course, my world record plane and the distance winner from Red Bull are in here too. If you're interested in officially knocking over my world record, there's still a thousand dollar reward for doing just that. Just use my design and officially break the Guinness World Record for distance, and the $1,000 is yours. You must be named by Guinness as the record holder.

The folding techniques get more complicated as you move through the book. The easy planes are first, and the folding gets crazy by the end. It's all step by step photos with multiple shots of the really hard stuff. The pictures of the finished planes show the exact way this plane will fly best. Try to match that photo for your fine adjustments.

The flight theory chapter comes at the end of the book, so any time you finish folding for the day, give those pages a look.

Paper:
All the planes here (except the Follow Foils and World Record Plane) are designed to work well with regular US printer paper -- 20lb 8.5x11 inches. I sometimes fold these planes with 24lb paper if they will be travelling with me or doing multiple shows in a day. 26lb paper is too thick for most of the collection.

For Follow Foils, I find phone book paper or 9lb onion skin paper works well.

For official world record competition, I use the maximum weight (100gsm A4) by Guinness rules. The smoothest, stiffest, paper stock I could find was Conqueror Paper CX22 Diamond White, unwatermarked.

Folding symbols and how to read the gauges are just over on the next page to the right; page 7.

Dart
Stunt
Glider
Easy
Advanced
Skill Level
Path

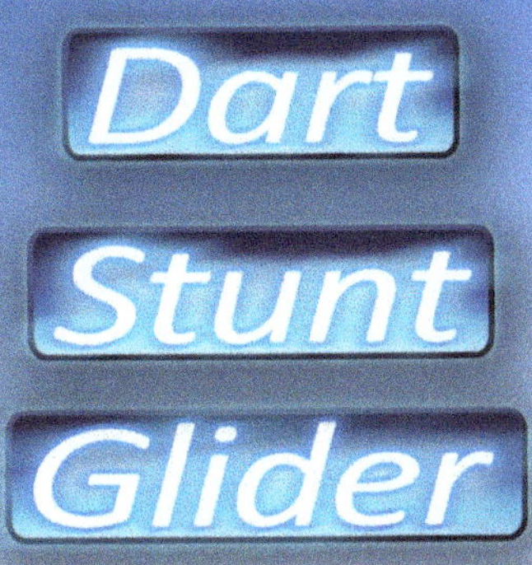

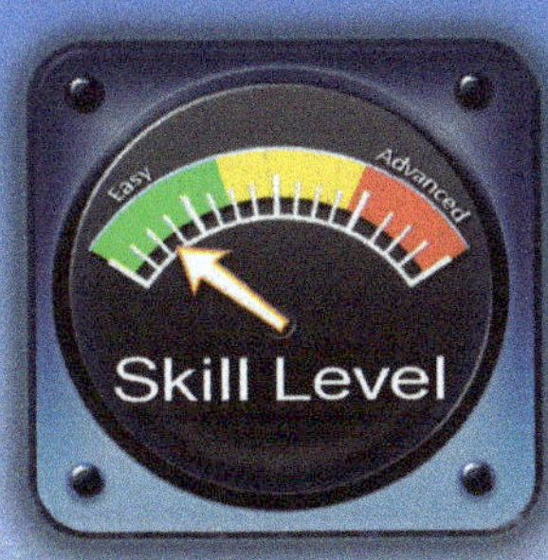

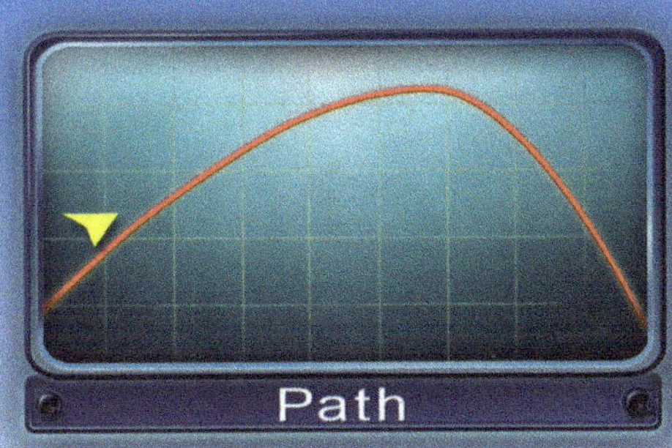

These gauges let you know what kind of plane, how easy it is to fold, and a typical flight path.

Folding Symbols

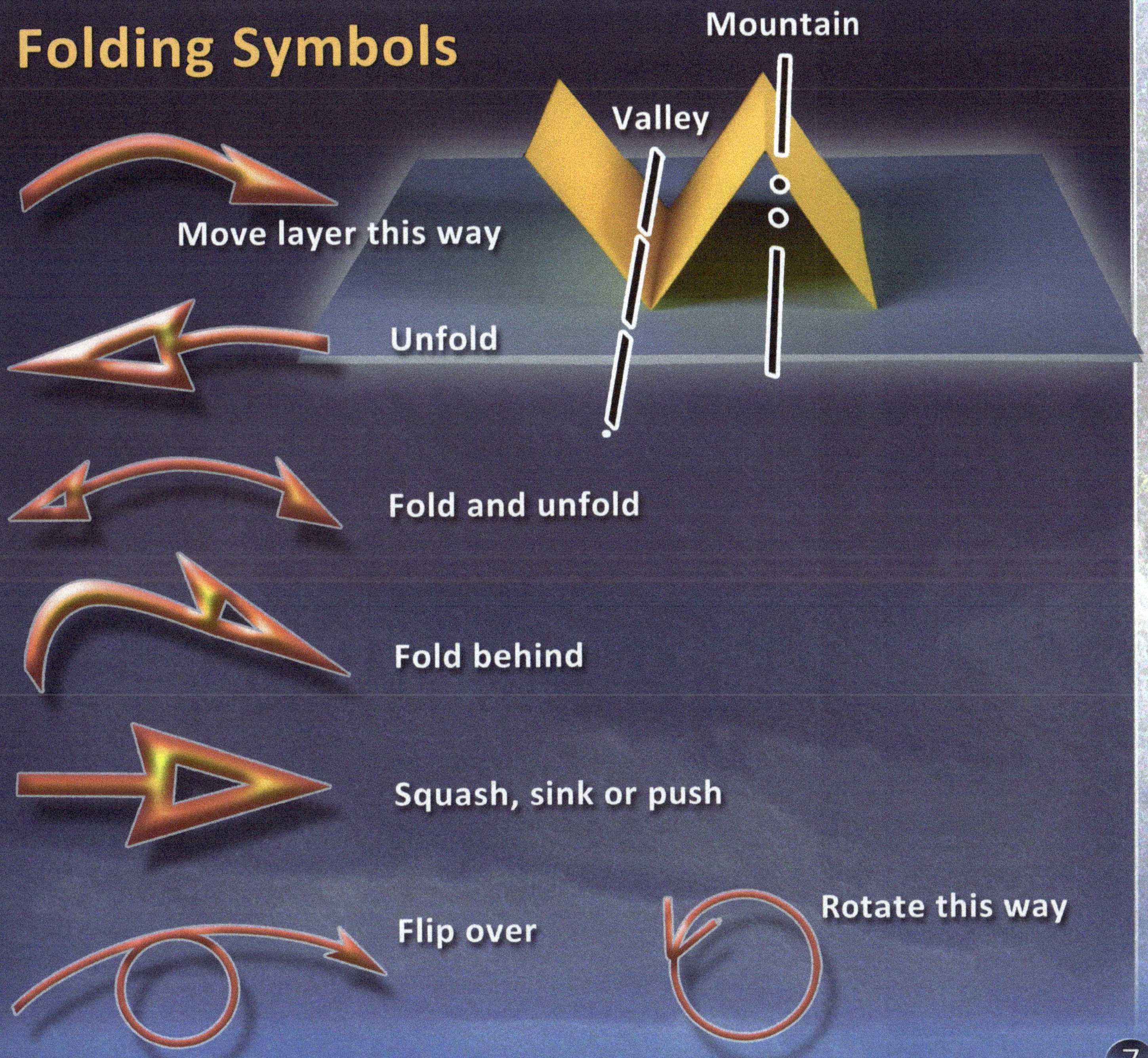

Mountain
Valley
Move layer this way
Unfold
Fold and unfold
Fold behind
Squash, sink or push
Flip over
Rotate this way

PHOENIX

Glider

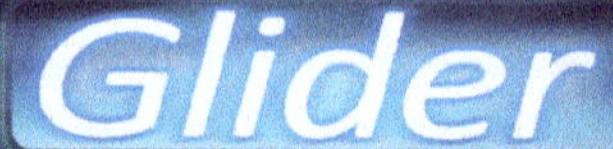

Easy to fold and great performance.
Fun indoors or outside.

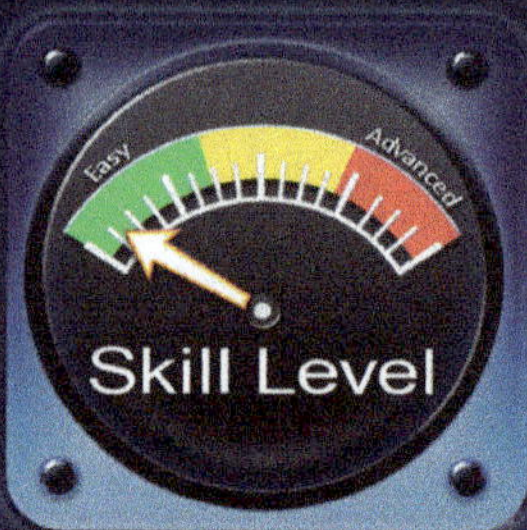

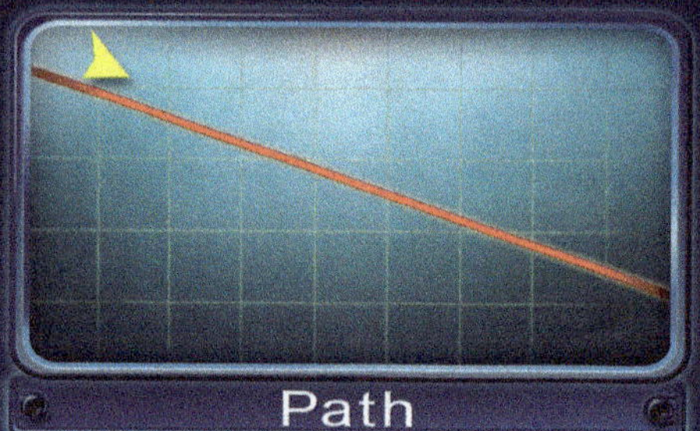

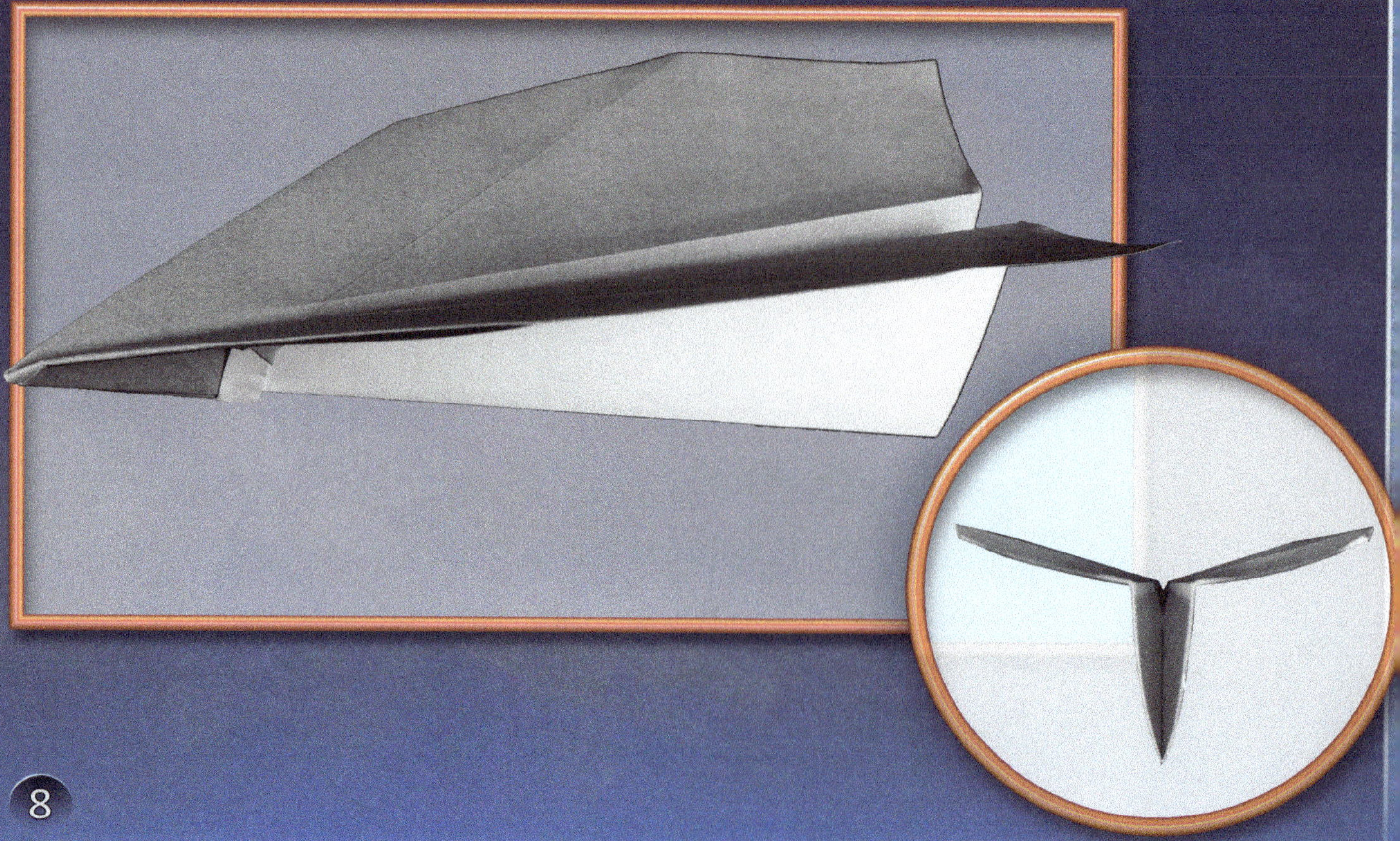

Phoenix Steps

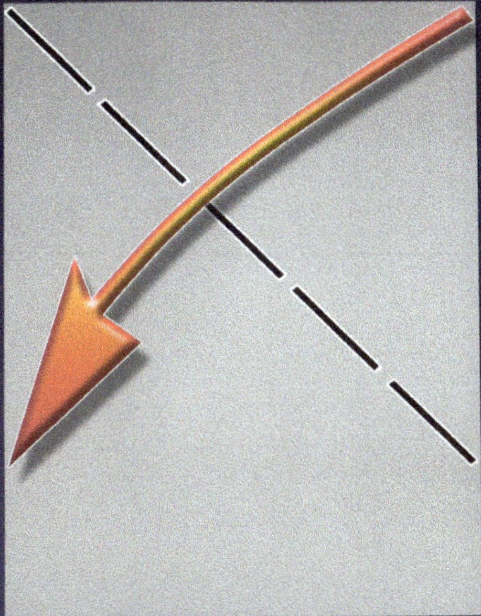

Make a diagonal fold. Line up the top with the left side.

Unfold step 1

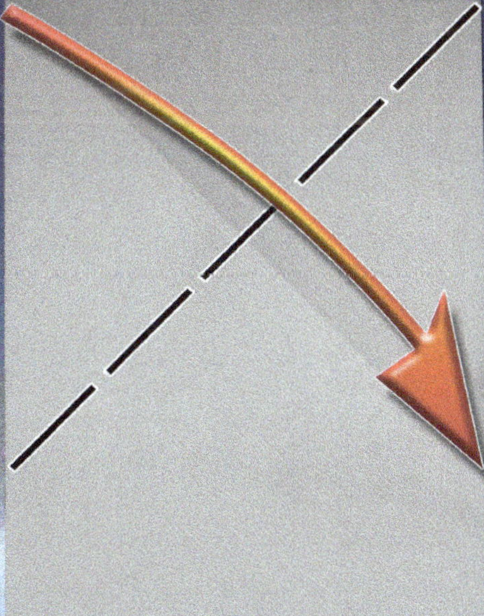

Make a diagonal fold the other way.

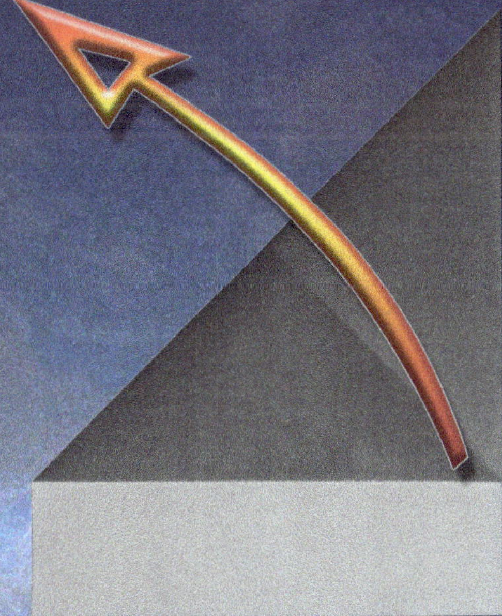

Unfold step 3

Phoenix Steps

5

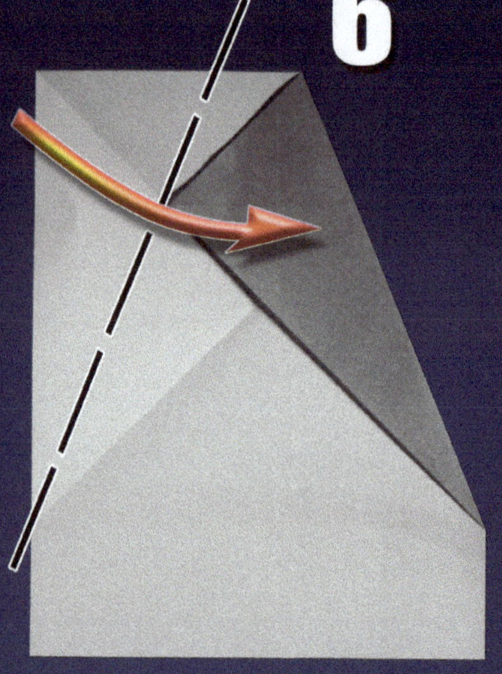

Fold the edge over to the diagonal crease.

6

Move the left side to the crease.

7

Fold the top down, using the existing crease.

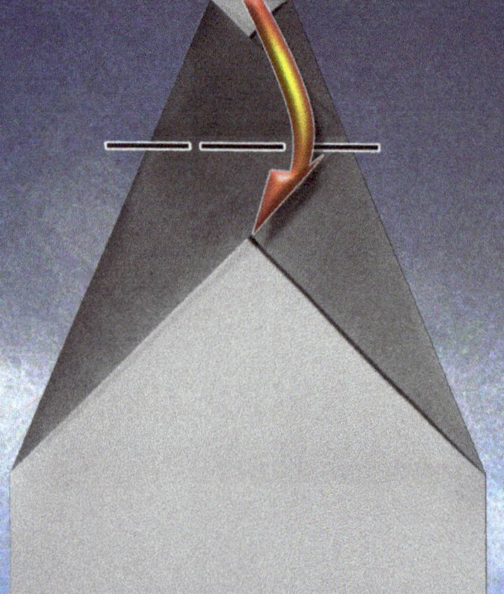

8

Flip the plane over.

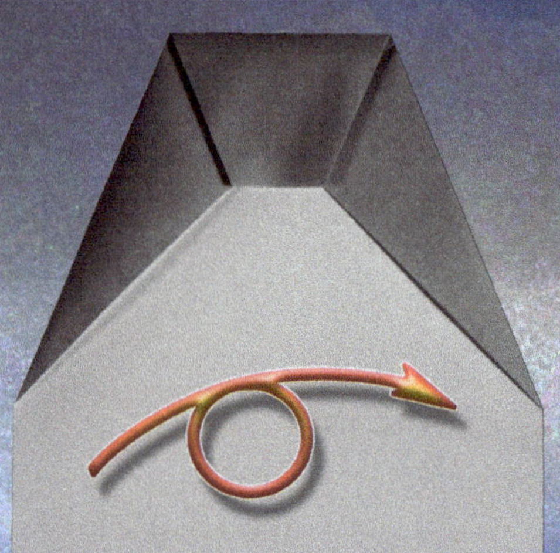

9

Rotate the plane
a quarter turn to the left.

10

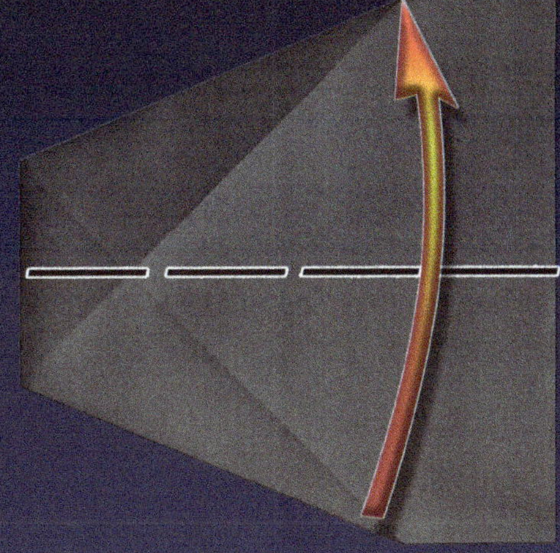

Fold the plane in half.

11

Fold the left edge down to line up
with the center crease.

12

Flip the plane over

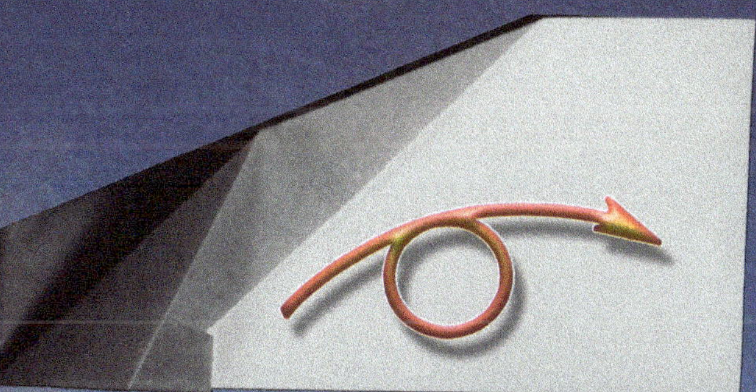

Folding Tip:
When moving multiple layers,
brace the bottom layers with your
thumbs to keep them from moving.

Phoenix Steps

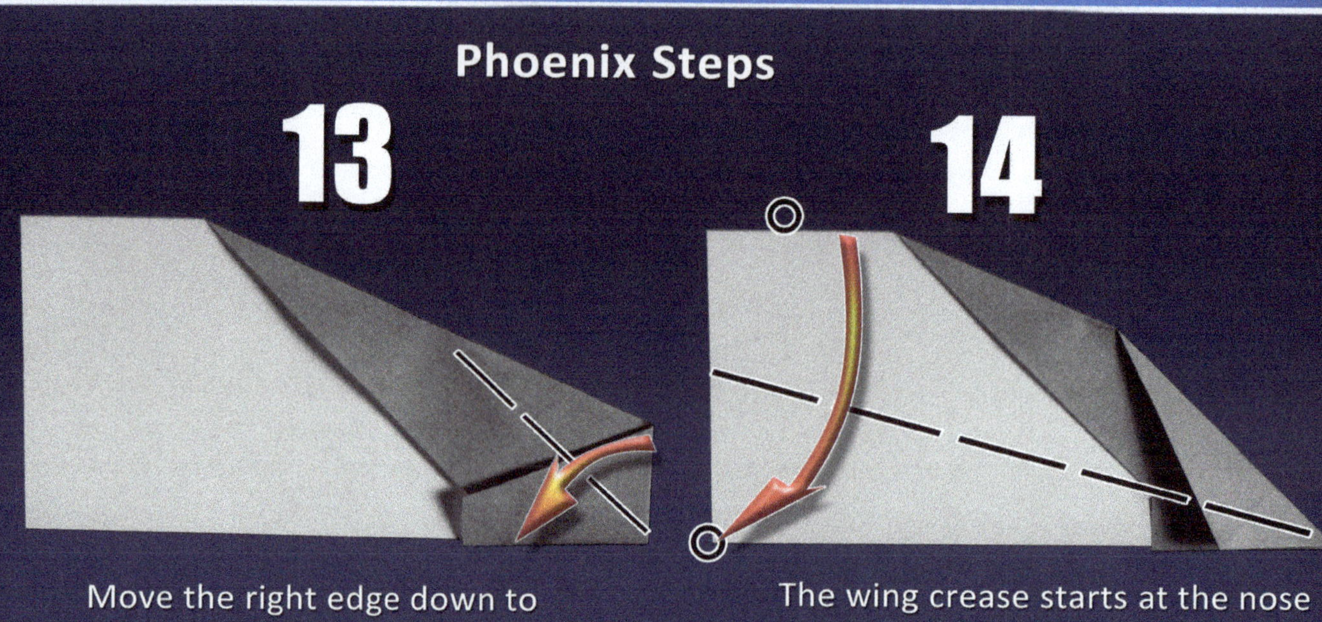

13

Move the right edge down to line up with the center crease.

14

The wing crease starts at the nose and slopes upward toward the tail. The raw edge of the wing will touch the corner of the tail.

15

Here's a closer look before you flip the plane over to make the other wing crease.

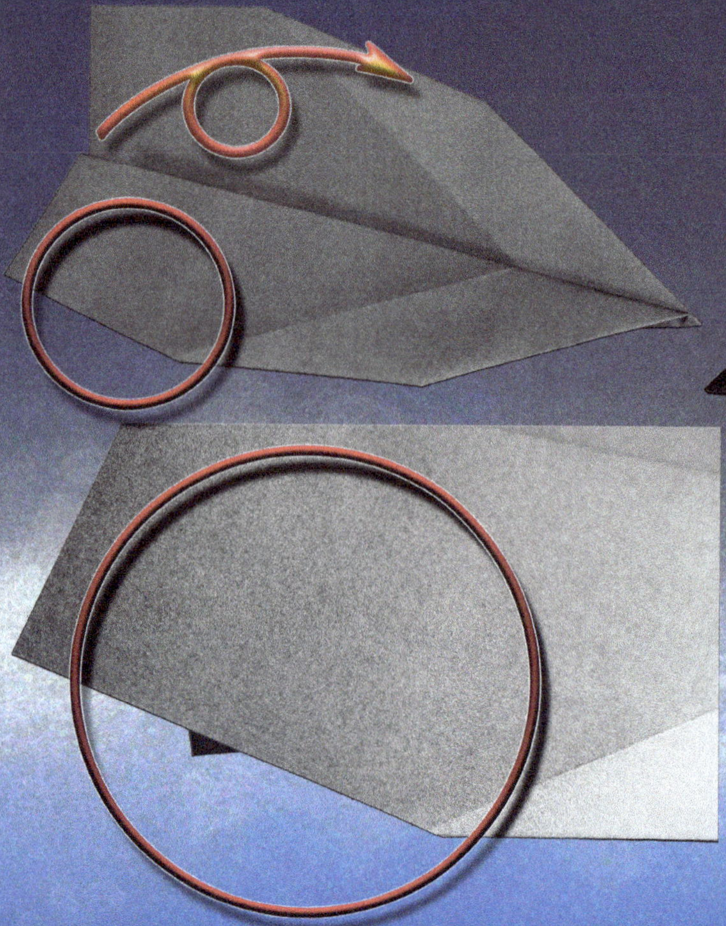

16

Make this wing match.

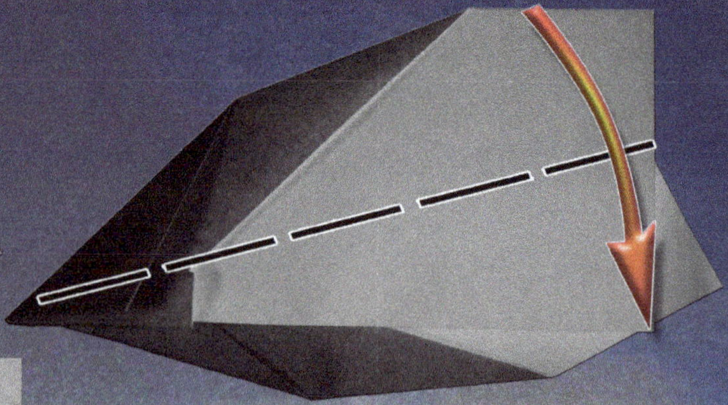

17

Open up the wings and center crease.

18

Fold the flap up.

19

Remake the wing and center creases.

SUZANNE

Glider

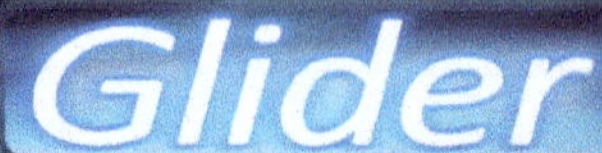

Current Guinness World Record Holder:
paper aircraft distance: 226 feet, 10 inches (69.14m)

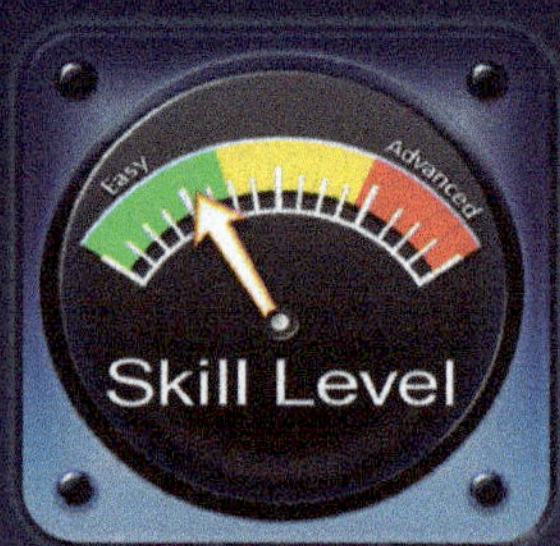

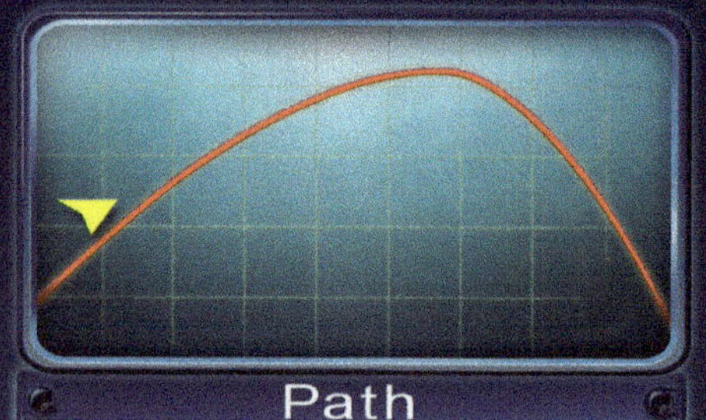

The folding is easy, but the taping will take some time.

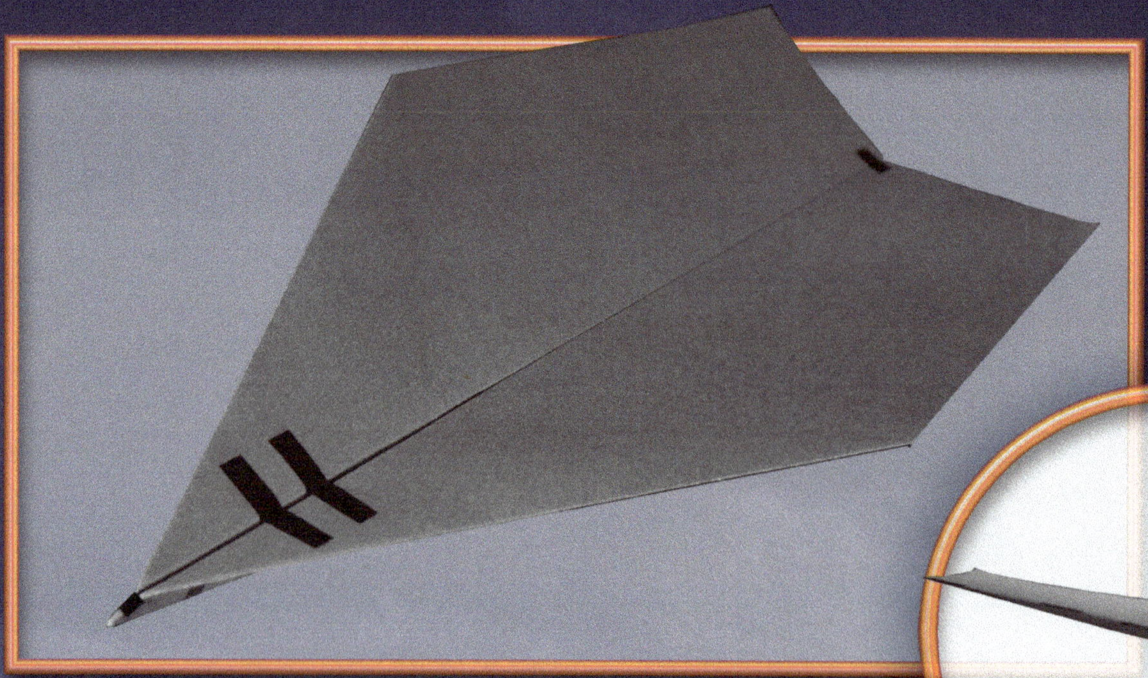

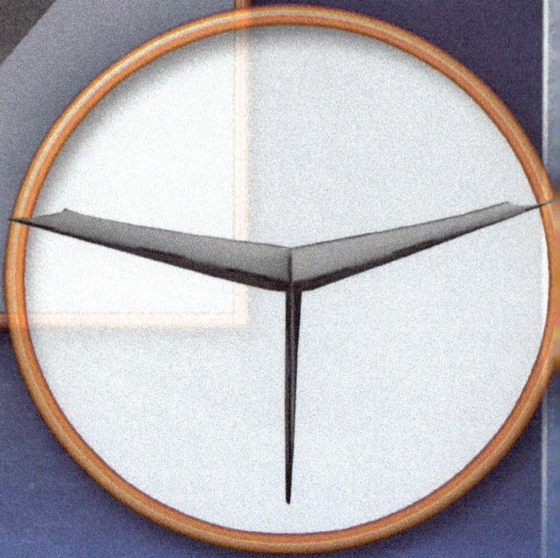

Getting Started

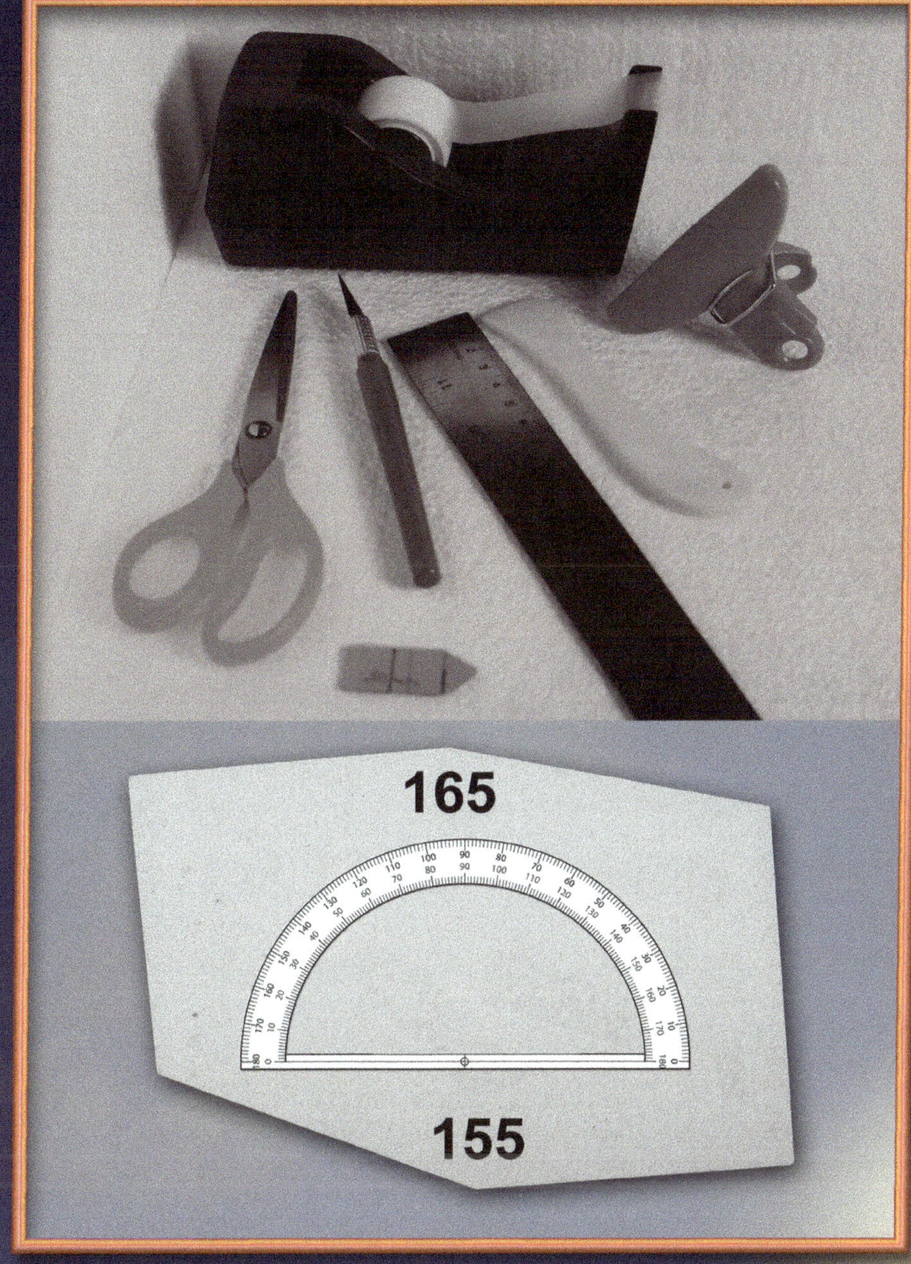

Tape: 25mm wide

Sharp Scissors

Hobby blade or chopstick for transferring tape

Ruler with millimeters

Folding Tool

Clip for holding the plane

Pre-cut measuring tool for creating 25x30 mm piece of tape

Protractor to create dihedral measuring tool

Cardboard dihedral tool created using the protractor. One side has a 165 degree angle, and the other side has a 155 degree angle.

Use an old shaving kit holder for your main tools. The angle gauge can stay with your paper.

I use Conqueror CX22, Diamond White, 100gsm A4 size paper for making world record planes.

If you don't have A4, remove 19mm from one side of a US letter size piece so that the paper looks taller. 26lb paper is the closest in thickness and weight.

Suzanne Steps

1

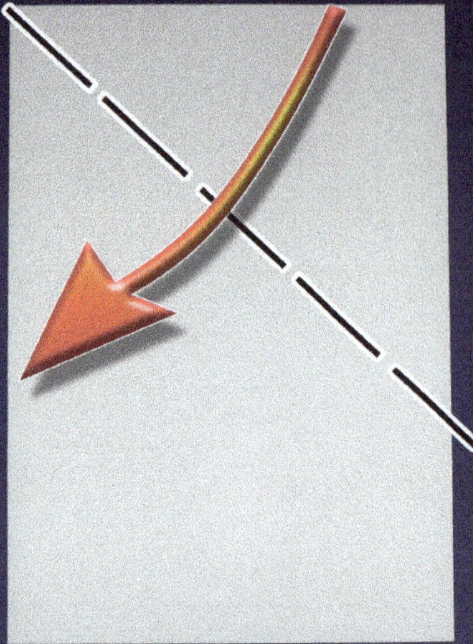

Make a diagonal fold. Line up the top with the left side.

2

Unfold step 1

3

Make a diagonal fold the other way.

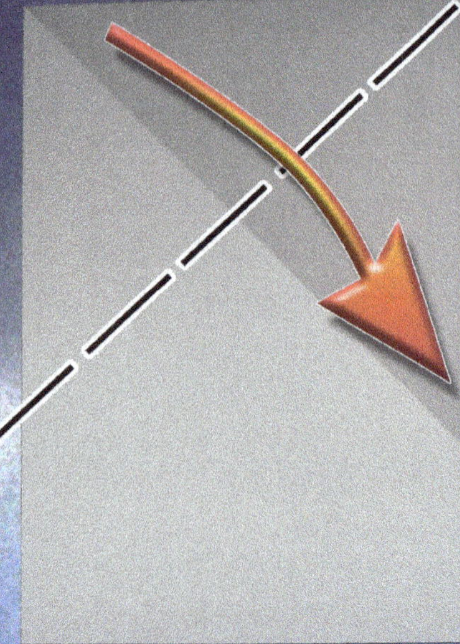

4

Unfold step 3

5

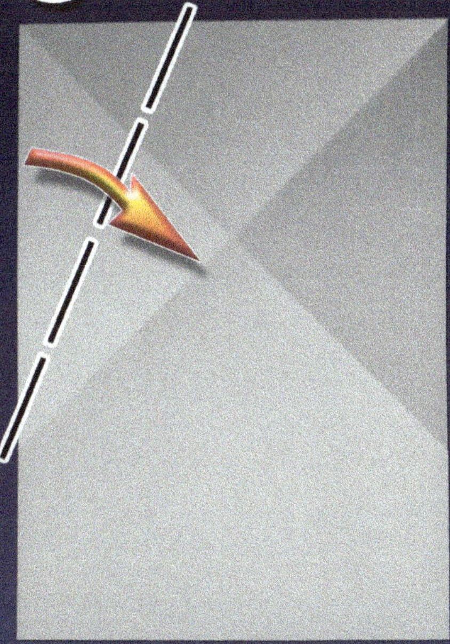

Line up the left edge
with the diagonal crease.

6

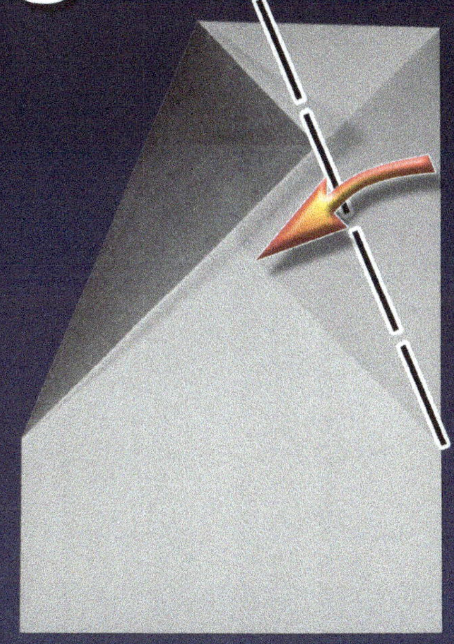

Line up the right side
with the diagonal crease.

7

Fold across the point where
the diagonal creases cross.

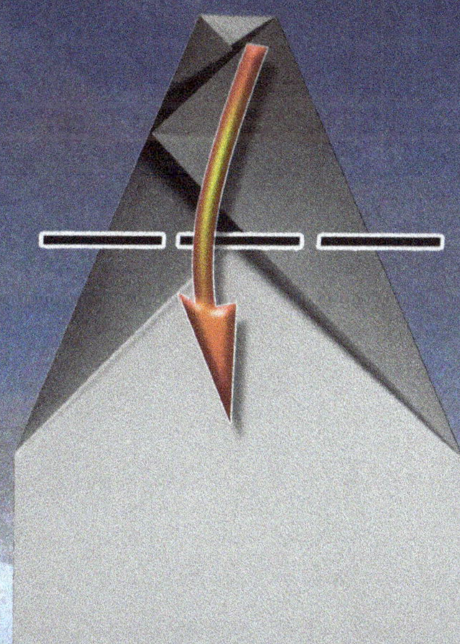

8

The top layer creases should
line up with the bottom layer.
Follow the existing creases.

Suzanne Steps

9

Fold the plane in half. The mountain fold means you flip it over first, and then fold in half.

10

1/4

Rotate it so that the center crease is on the bottom.

11

Make the wing fold. It's like The Phoenix. The next steps show another way to get there.

11a

Position the creased edge against the center crease. Don't Fold.

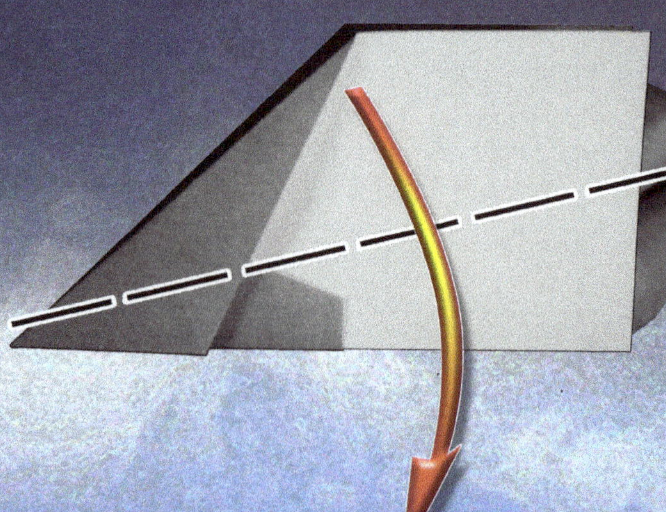

Keep moving the wing down until the white triangle is gone.

Suzanne Steps

11b

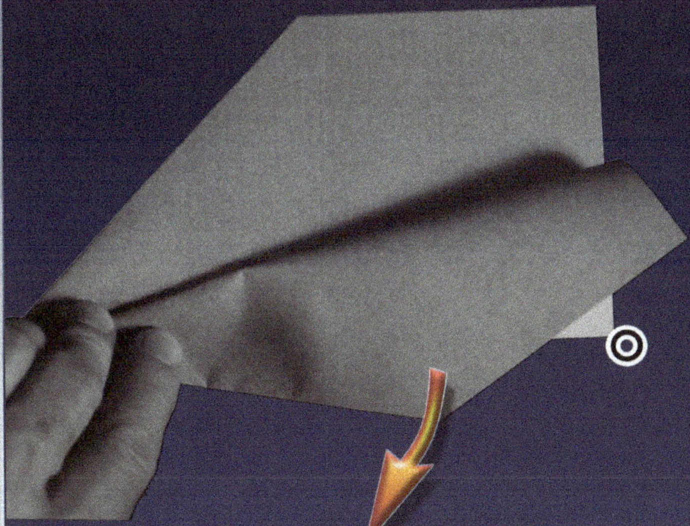

Keep pulling the wing down until the little triangle is gone.

11c

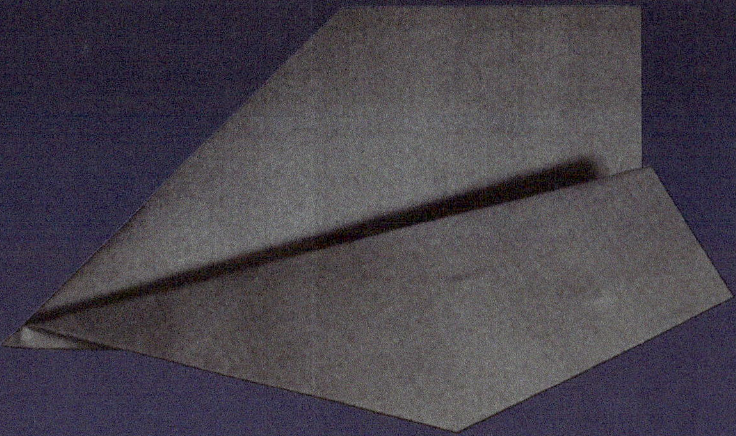

Make the other wing match.

12

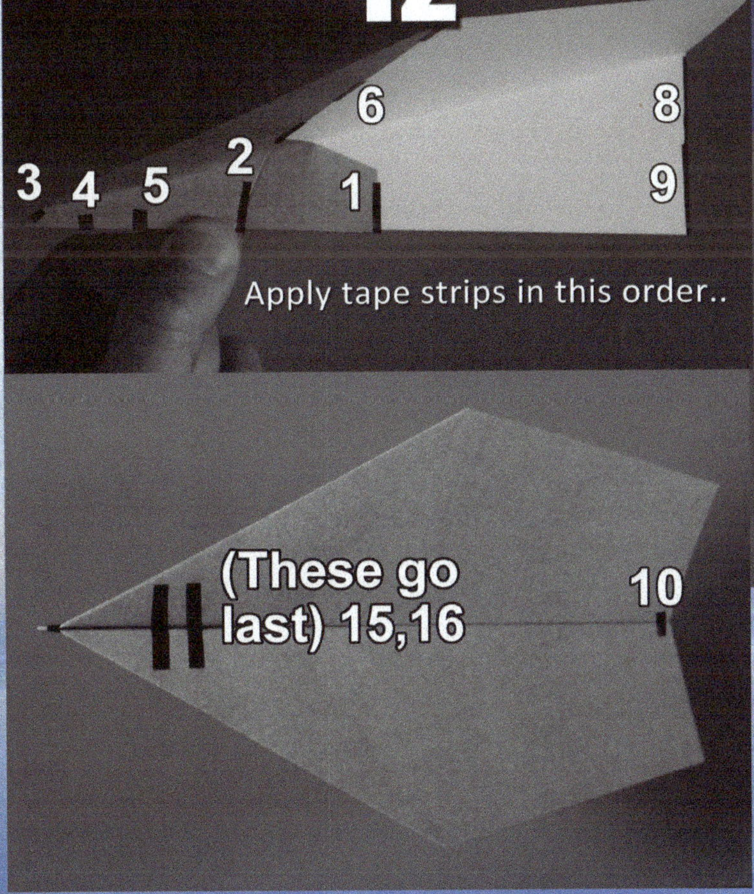

Apply tape strips in this order..

13

Note 3,4,5, 15 &16 are wider strips.
3,4 & 5 all come from one strip.

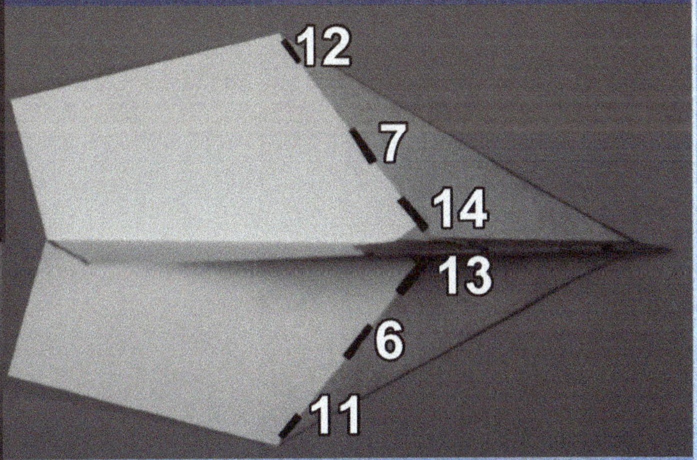

10, 11 & 12 are from one strip.

14

Use your dihedral gauges to set the angle at the nose at 165 degrees. At mid-wing 155 degrees.

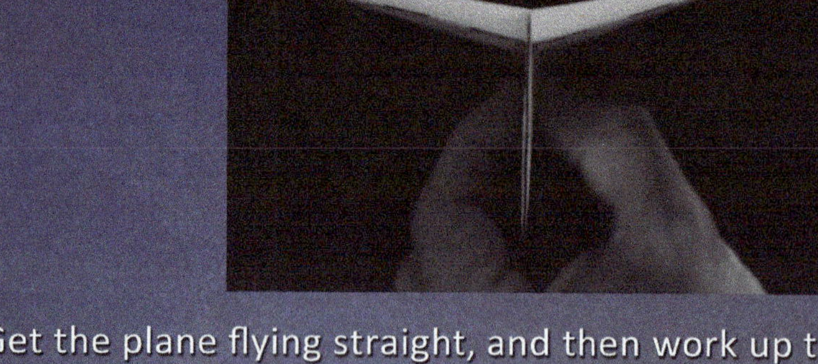

$1,000 Reward:
Use this design and officially break the world record for distance to win $1,000. You must be named by Guinness as the world record holder.

Get the plane flying straight, and then work up to a fast throw. Keep in mind that the faster the plane flies, the closer to the nose your adjustments will be made. At our highest throwing speeds, if the plane veered right on launch, I would bend the edge on the left down a little, at the nose. For turns at regular gliding speeds, adjust the rear of the plane.

Throwing Tips:
▶ Hold where the most layers meet (thickest part)
▶ Keep the wings level
▶ Smooth acceleration

Use these tips for most planes.

THE PLANE

Glider

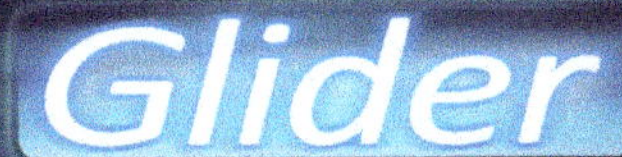

The Plane is a reliable, high performance paper airplane for indoors and outdoors.

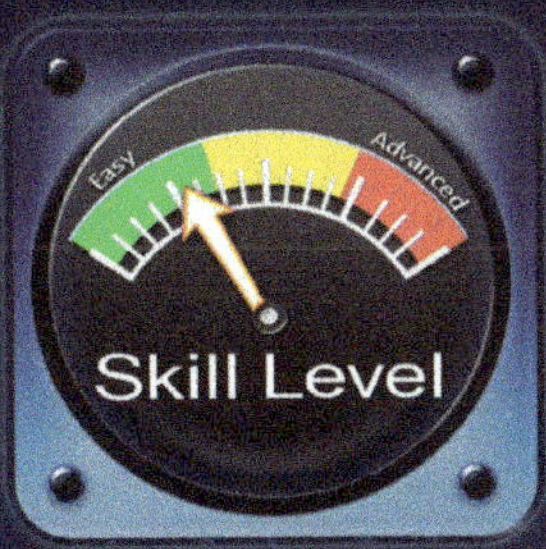

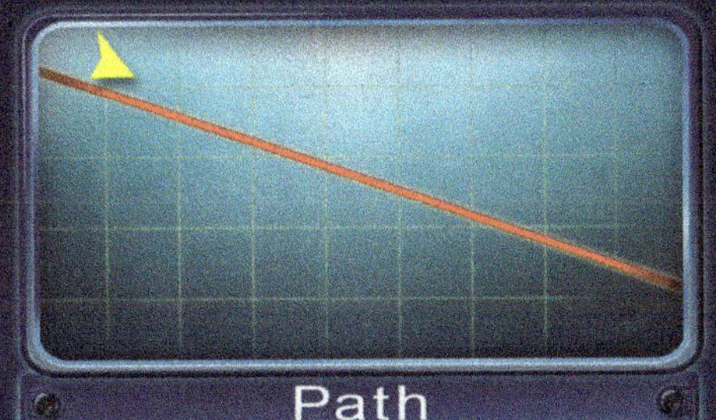

As with most gliders, add more up elevator for outdoor fun.

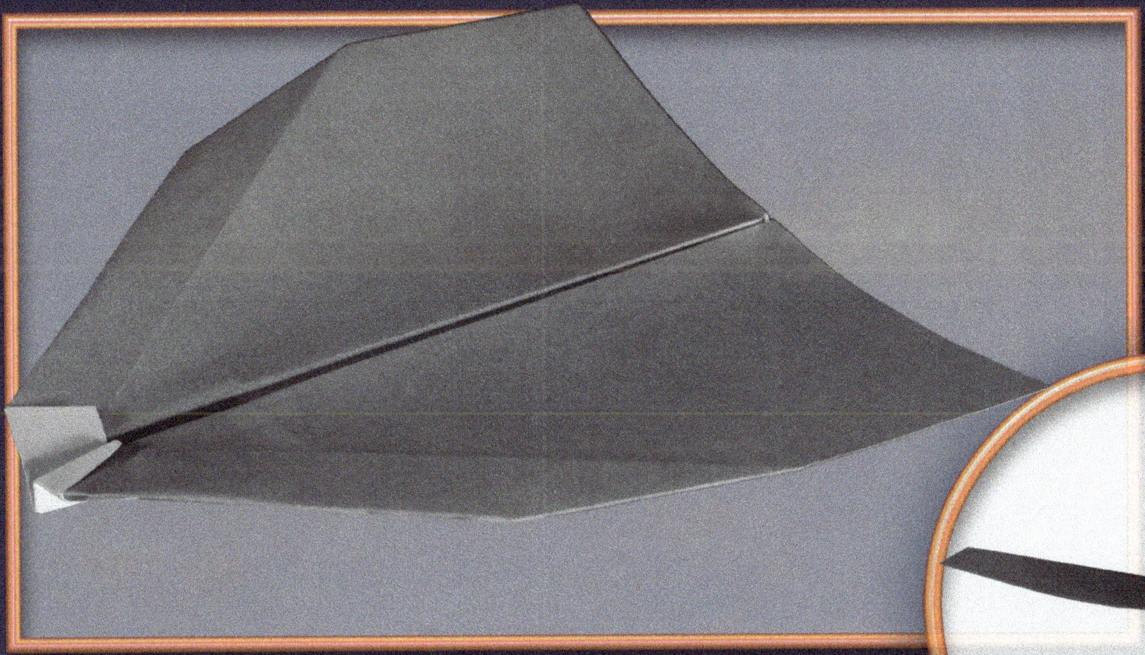

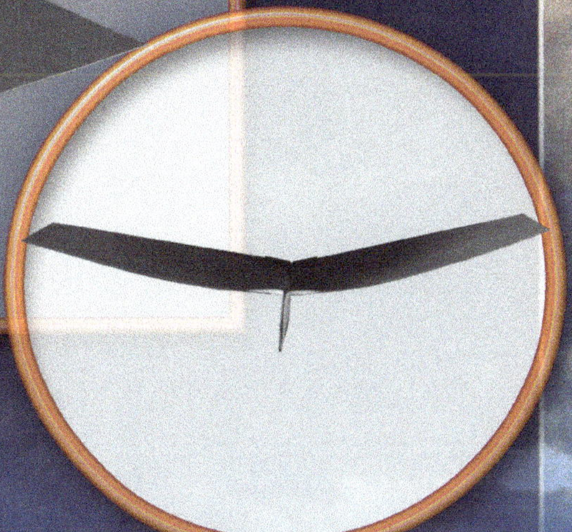

The Plane Steps

1

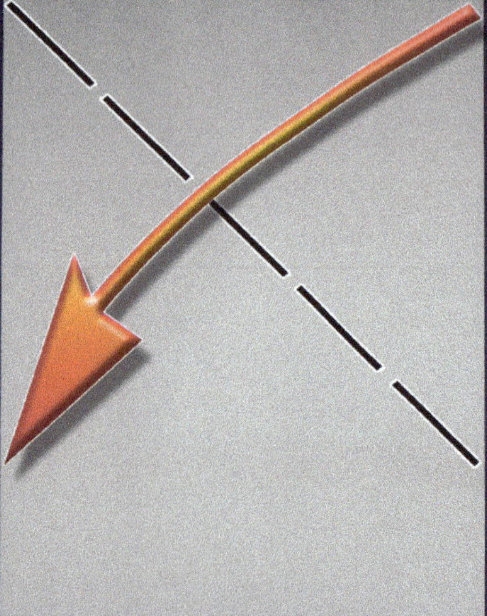

Make a diagonal fold. Line up the top with the left side.

2

Unfold step 1

3

Make a diagonal fold the other way.

4

Unfold step 3

5

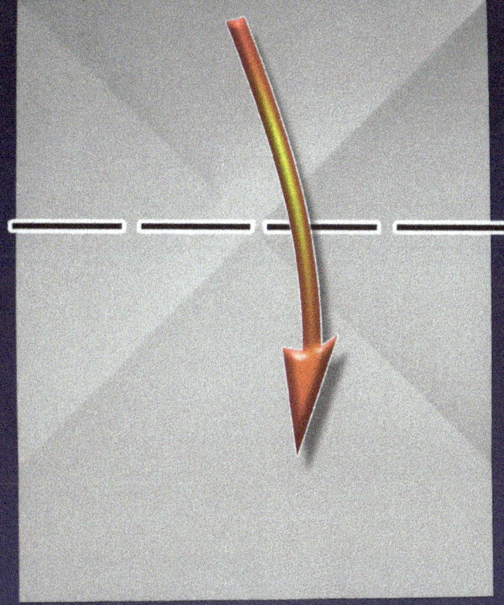

Fold across the center of the diagonal folds.

6

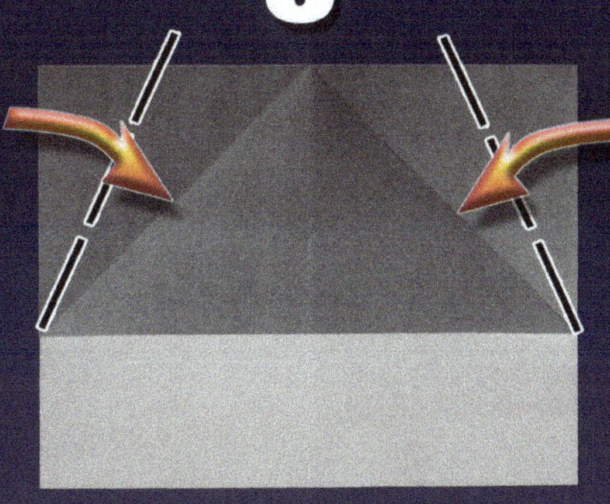

Fold the outside flaps to the diagonal creases. A reverse fold is coming.

7

Unfold step 6

8

Reverse the corners.

9

The lower right marked corner moves up to touch the marked crease.
Use the marked upper right corner as one end of the fold.

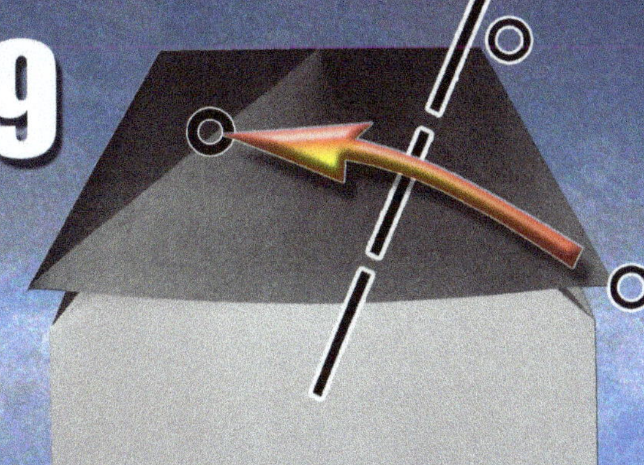

10

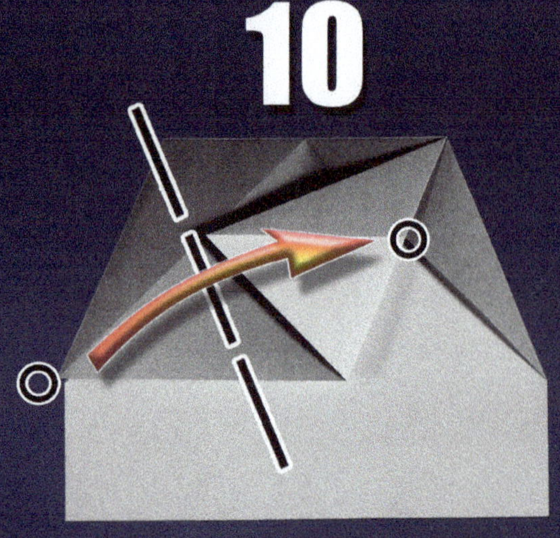

See how the edge of the gray layer, on the flap we just moved, is parallel with the top edge? That's a quick check that you made step 9 correctly. Repeating 9 for the left side is easier. Just move the marked point to the marked corner.

11

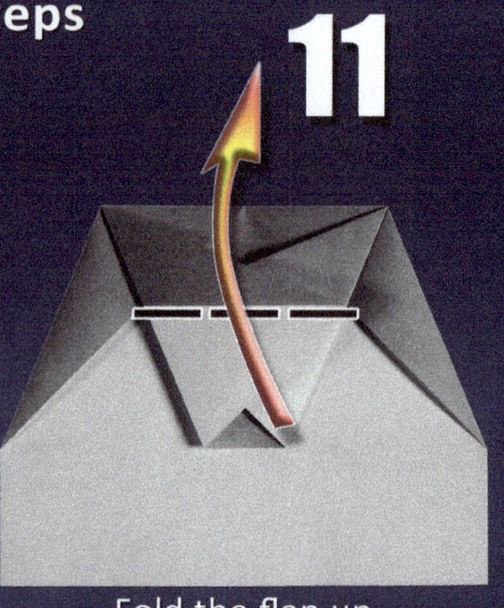

Fold the flap up.

12

Flip the plane over.

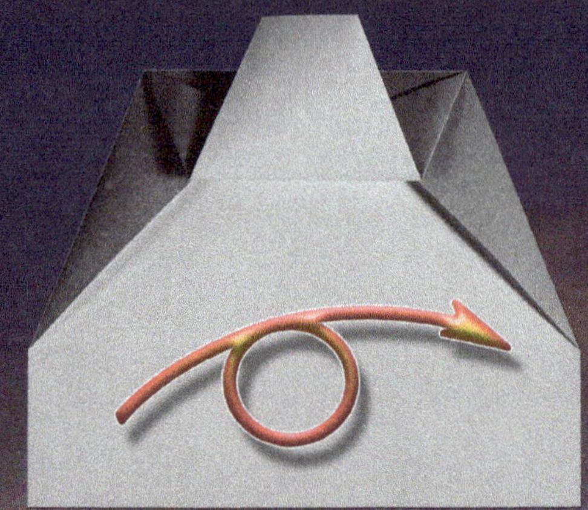

13

Fold the flap down.

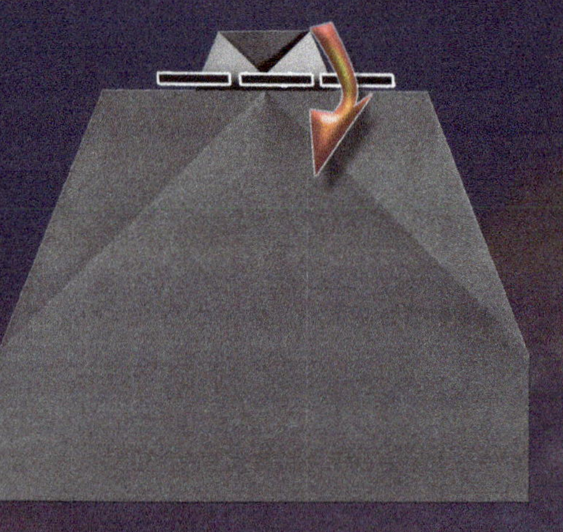

14

Rotate the plane a quarter turn right.

15

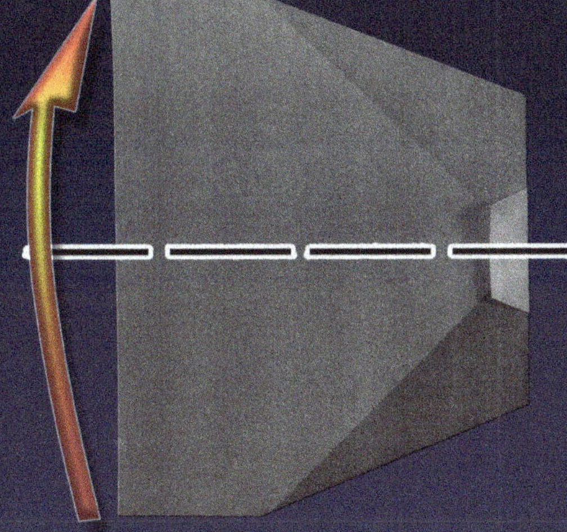

Fold the plane in half.

16

Tuck the upper right corner into the pocket. Do the same for the other side.

17

The wing fold starts about half way up the short edge at the nose. The tail should be no taller than the short edge at the nose.

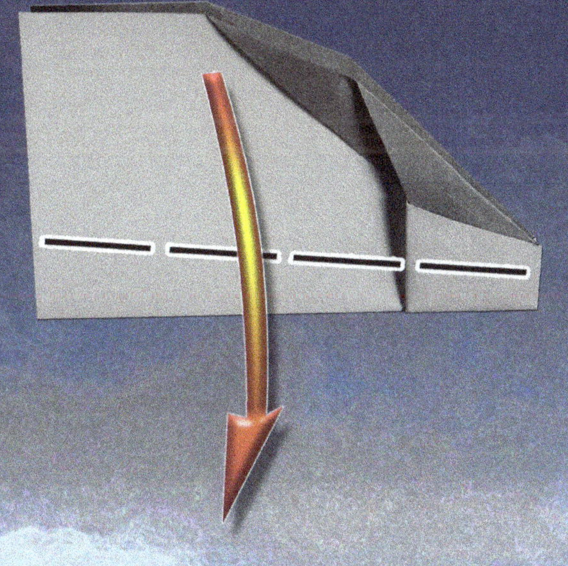

18

Flip the plane over and make the other wing match. Take care to line up the rear corners of the wings. The middle corners can be a smidge off.

The Plane Steps

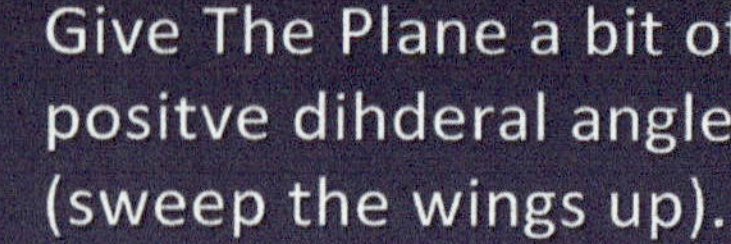

Give The Plane a bit of positve dihderal angle (sweep the wings up).

Notice the small bends at the tail. That's enough up elevator to keep The Plane flying indoors. You may need more outdoors.

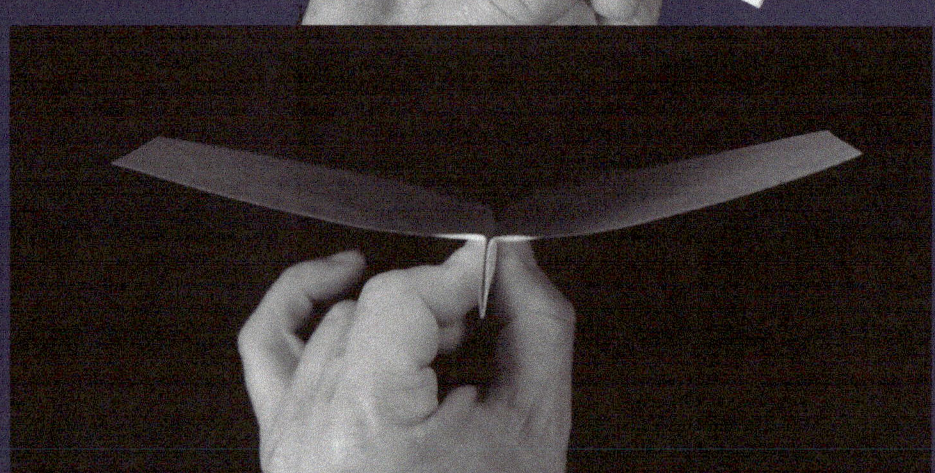

Fly a glider slowly at first to see if it tracks straight in a regular glide. Work up to harder throws. We used that method practicing for, and breaking the record.

STEALTH

Dart

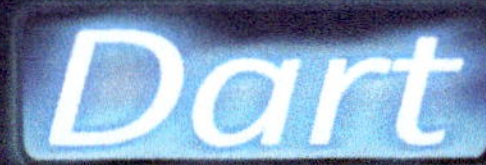

Muscle, style, performance, and so under the radar; there's no way to tell what hit you.

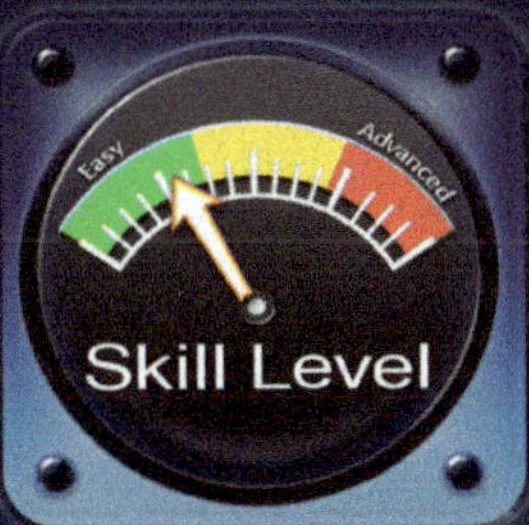

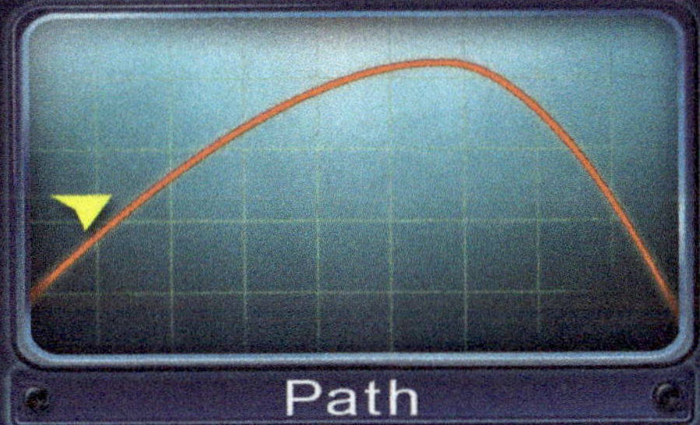

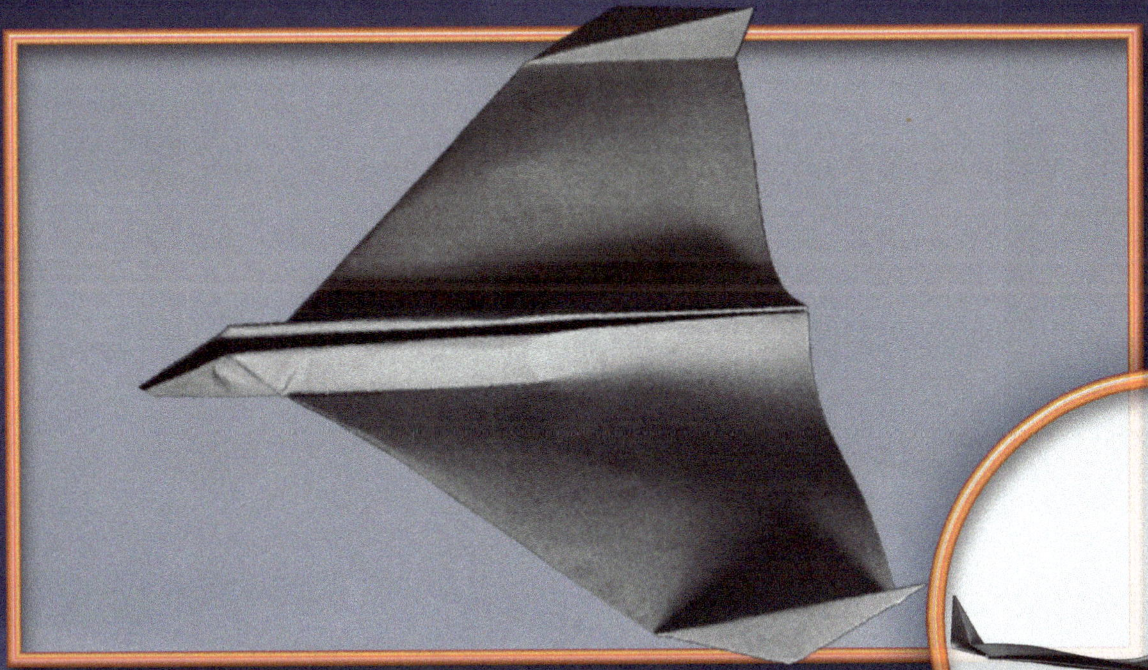

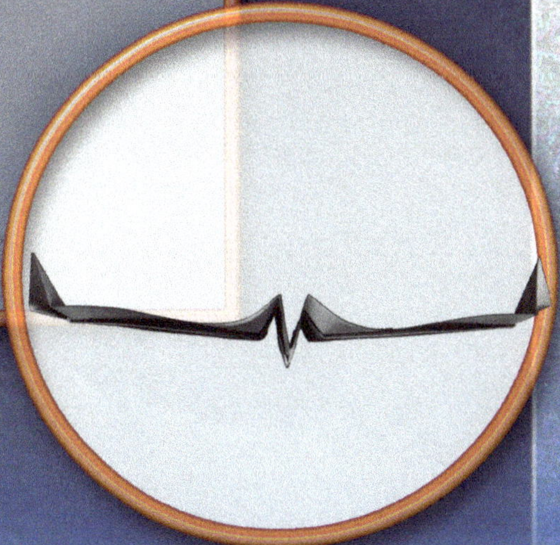

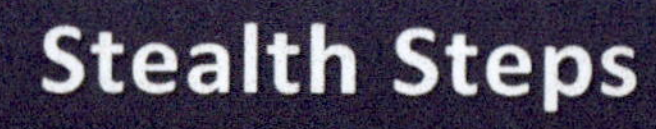

Stealth Steps

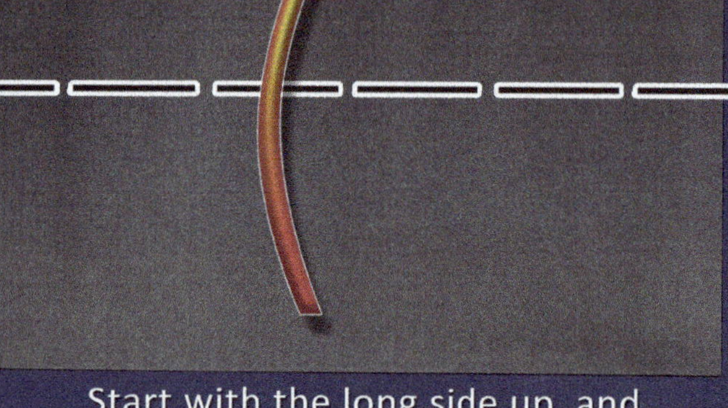

Start with the long side up, and fold the page in half.

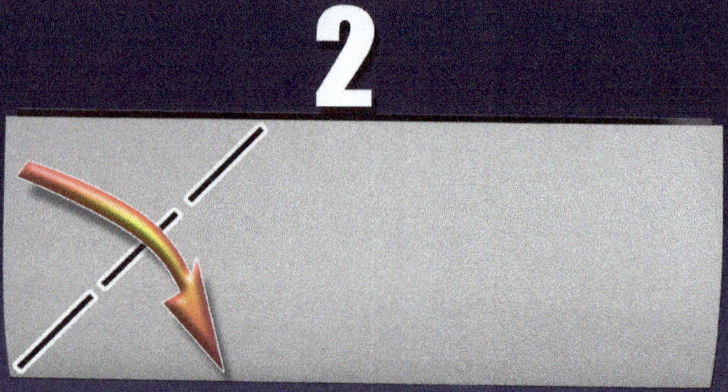

Fold the corner down. The top layer only. Do the same to the other side.

3

Unfold the center crease and rotate so the point is on top.

4

Fold the top to the bottom.

Tips on folding:
Careful folding is the key for all great paper airplanes. Small errors at the start add up to big errors at the end. Start smart.

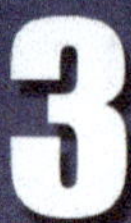

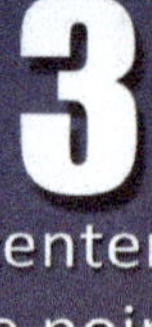

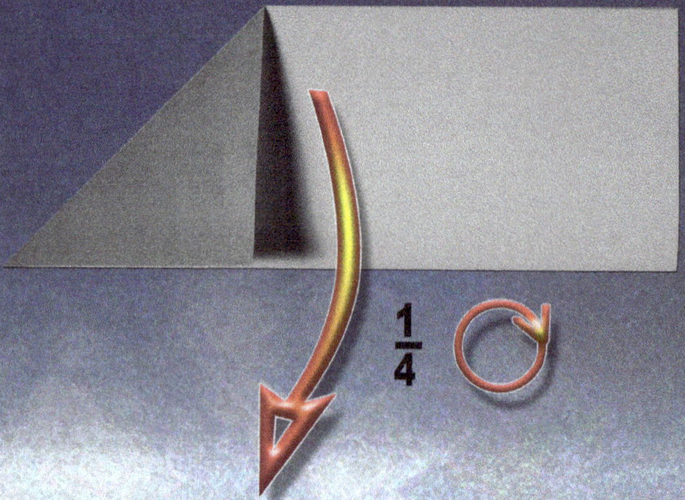

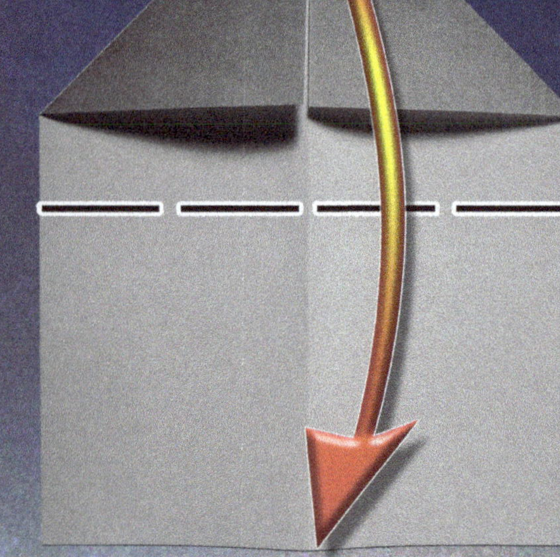

5

Fold the point back to the center of the top.

6

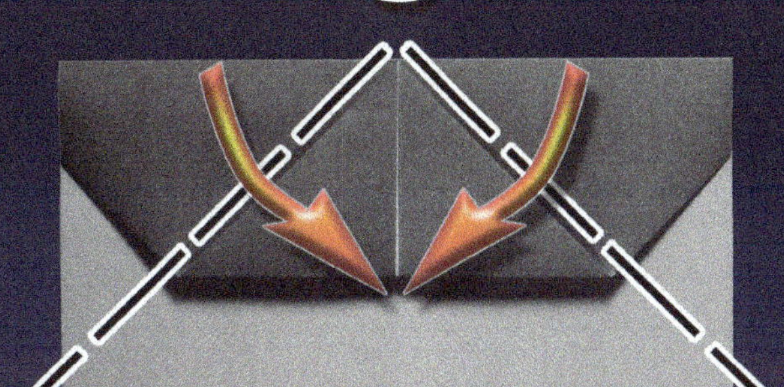

Fold the corners down to the center crease.

7

Fold from the bottom of the Step 5 crease to the outside corner.

8

Use the Step 7 crease as a guide to tuck the flap under.

9

Fold the right side under. This shows how the mountain fold symbol looks.

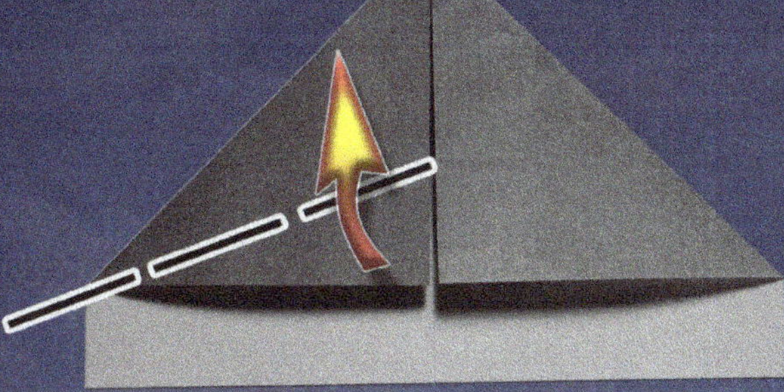

10

Fold down the center, making a mountain fold. Rotate left.

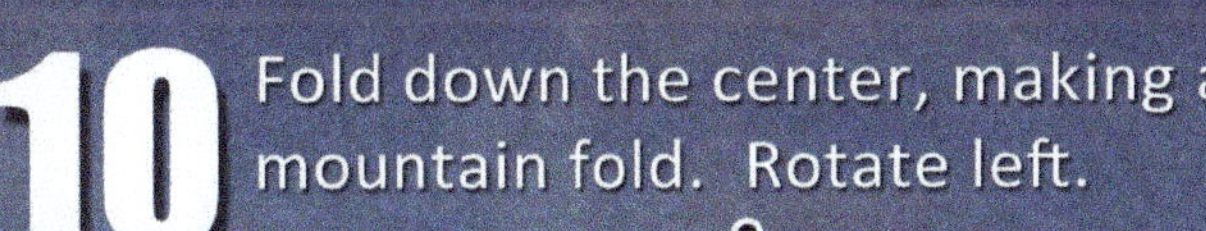

$\frac{1}{4}$

Stealth Steps

11

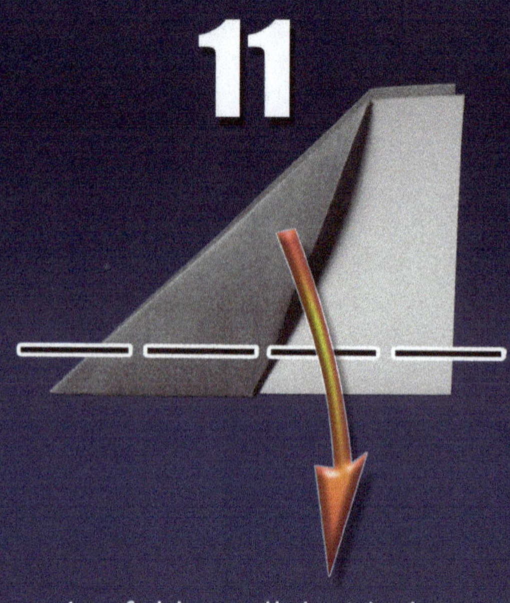

Make a wing fold parallel with the center crease, and an inch tall (22mm).

12

Make the other wing match.

13

Fold up the wing tip about the same height as the body (22mm).

14

Make a diagonal fold across the winglet from step 13

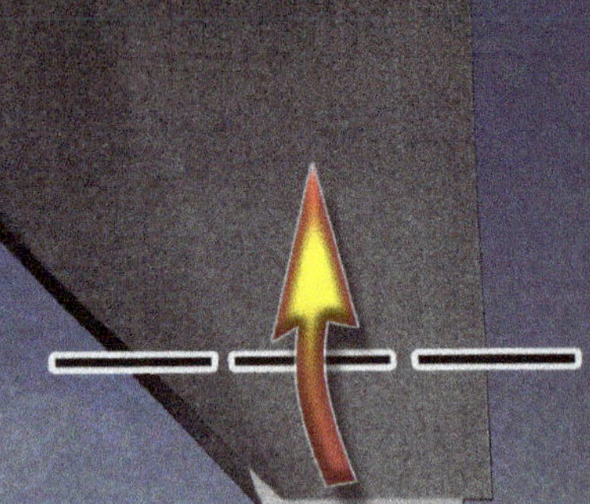

15

Repeat steps 13 and 14 for the other wing.

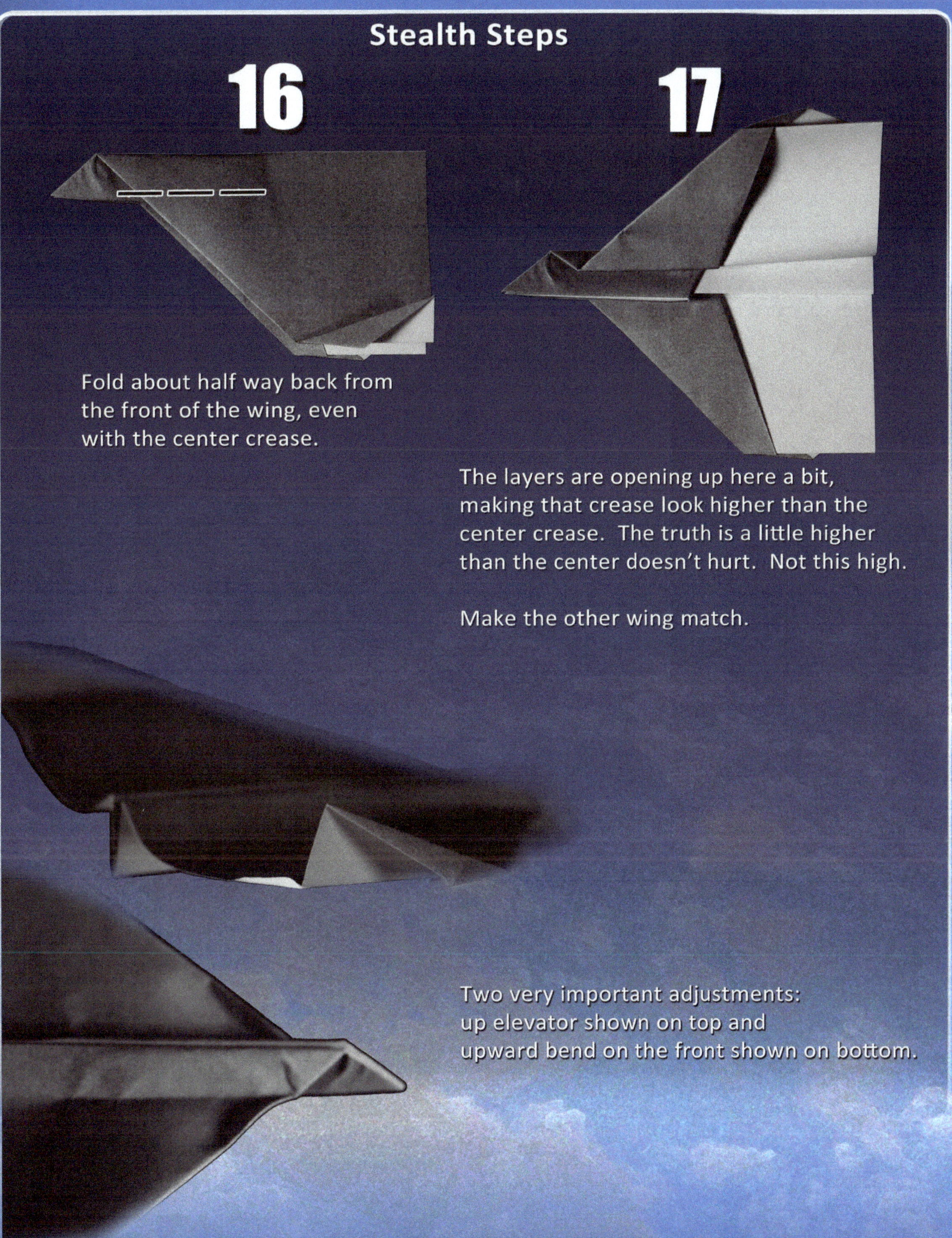

16

Fold about half way back from the front of the wing, even with the center crease.

17

The layers are opening up here a bit, making that crease look higher than the center crease. The truth is a little higher than the center doesn't hurt. Not this high.

Make the other wing match.

Two very important adjustments:
up elevator shown on top and
upward bend on the front shown on bottom.

PHAT GLIDER

Stunt

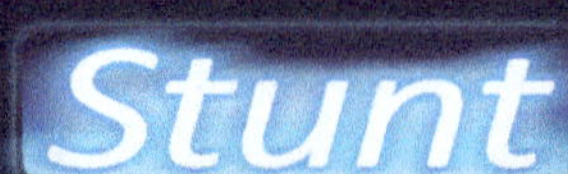

A great stunt plane: circles, loops, and inverted flights are all packed into an easy to fold plane.

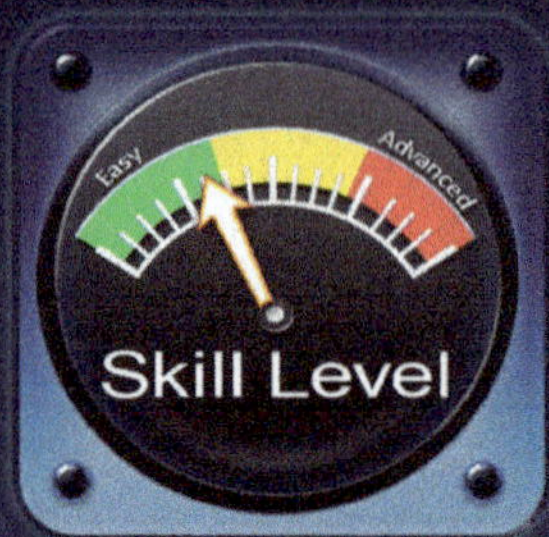

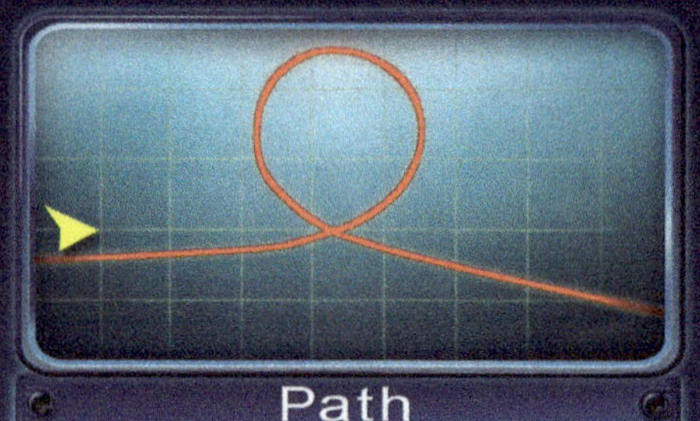

Throwing is a big key to stunts. Remember to tilt the plane over for circles.

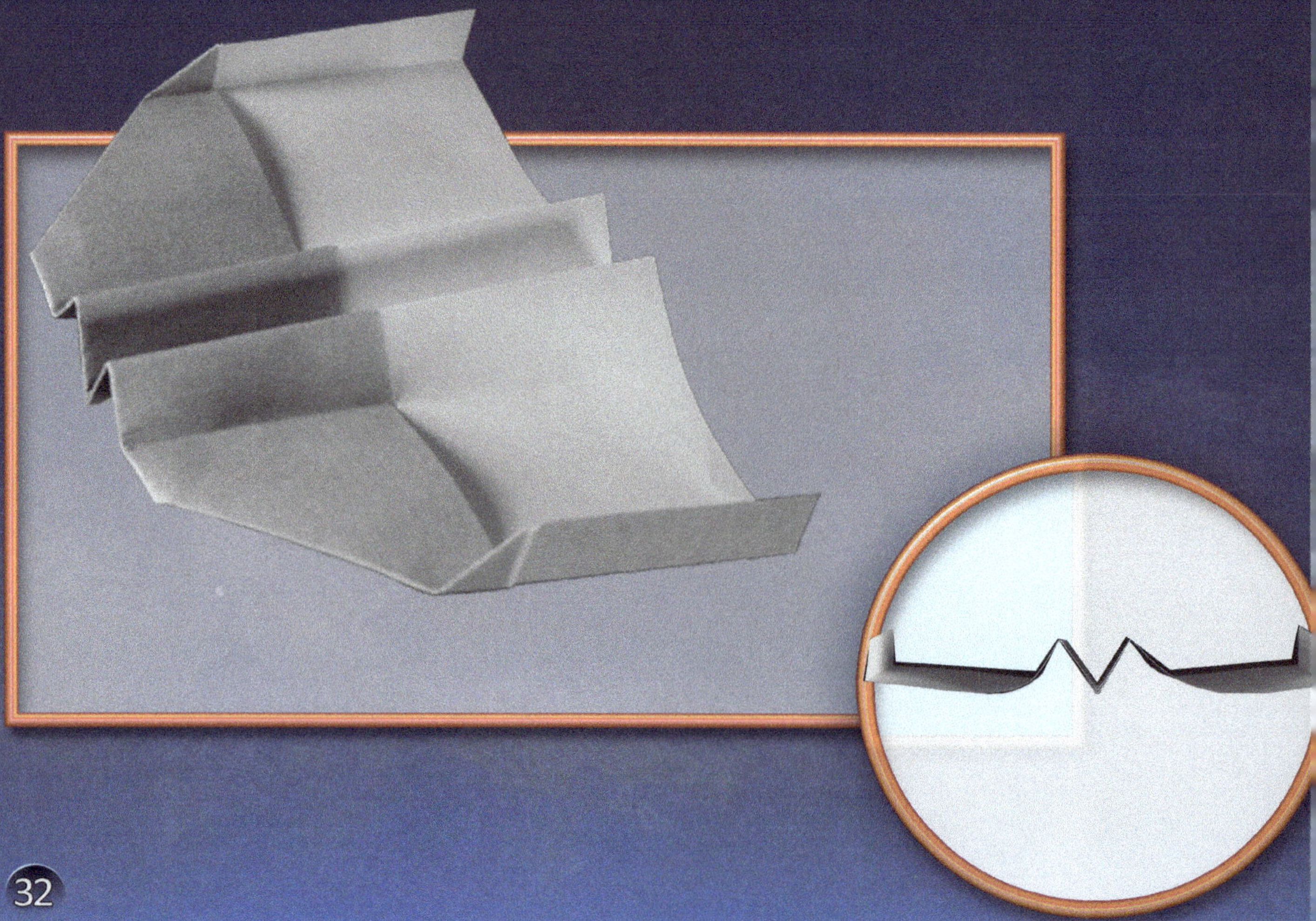

Phat Glider Steps

1

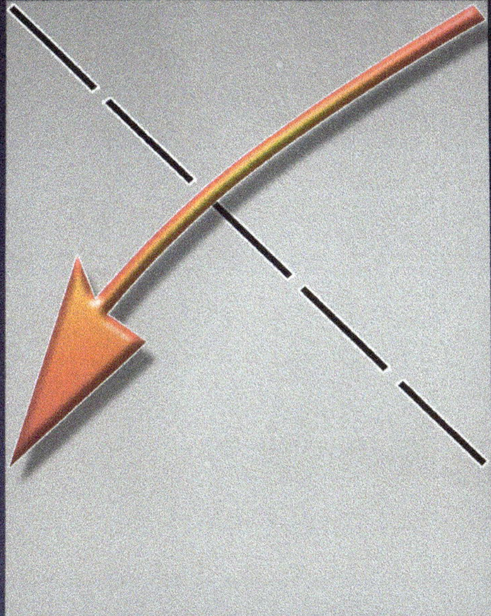

Make a diagonal fold. Line up the top with the left side.

2

Unfold step 1

3

Make a diagonal fold the other way.

4

Unfold step 3

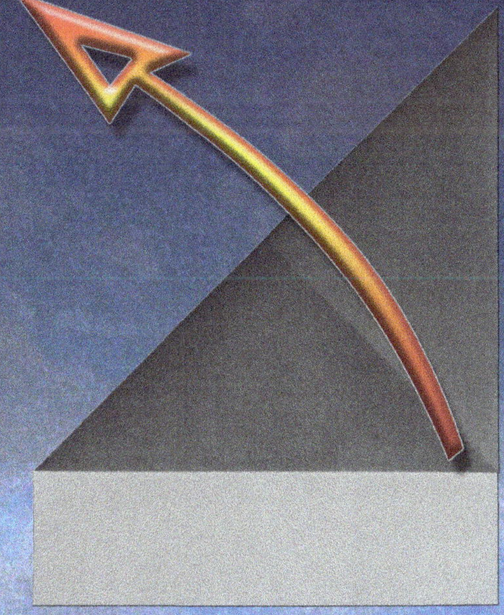

Phat Glider Steps

5

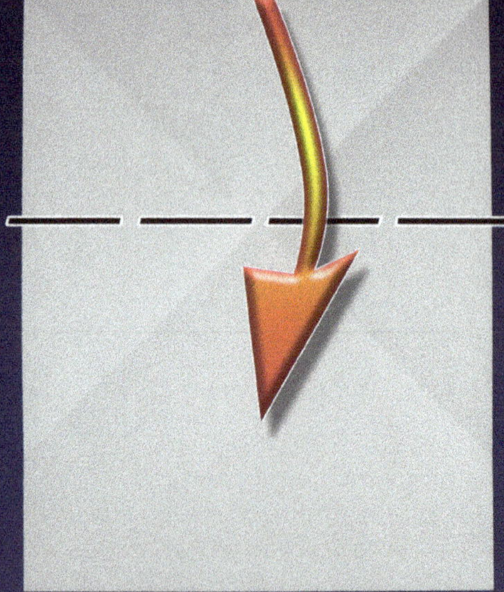

Fold the X in half.

6

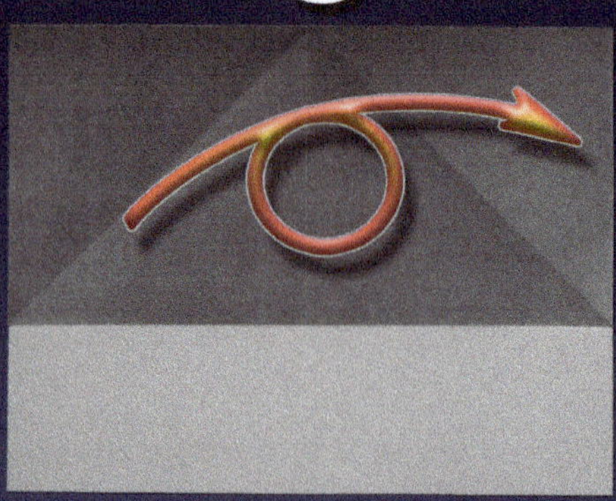

Flip the plane over.

7

Follow the existing creases
to move the corners down.

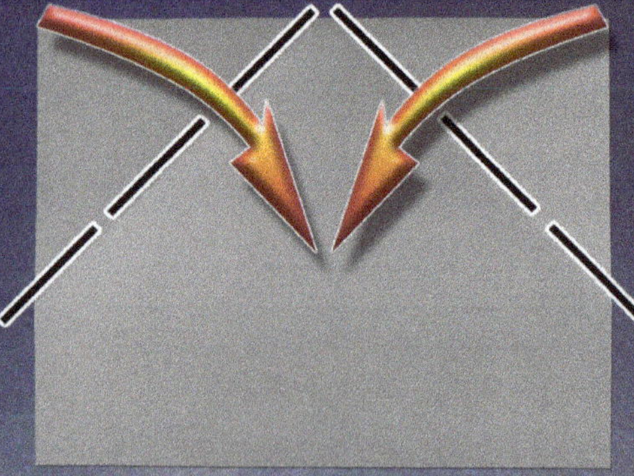

8

Fold the top
down to the
center.

10

$\frac{1}{4}$

Rotate the plane.

9

Flip the plane over.

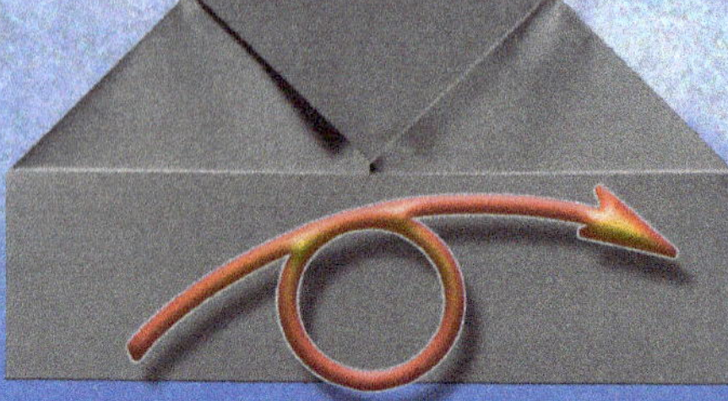

Phat Glider Steps

11

Fold the plane in half.

12

Make the wing crease. The front edge is folded in half. You can make the wing a tiny bit longer, but not any shorter.

13

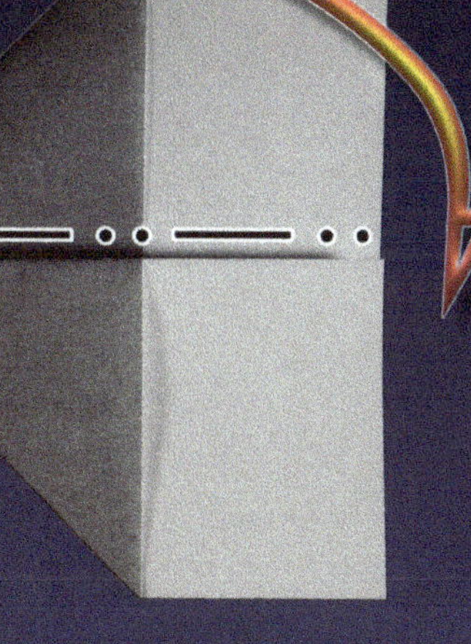

Make the other wing match.

14

Make a winglet about half the height of front edge.

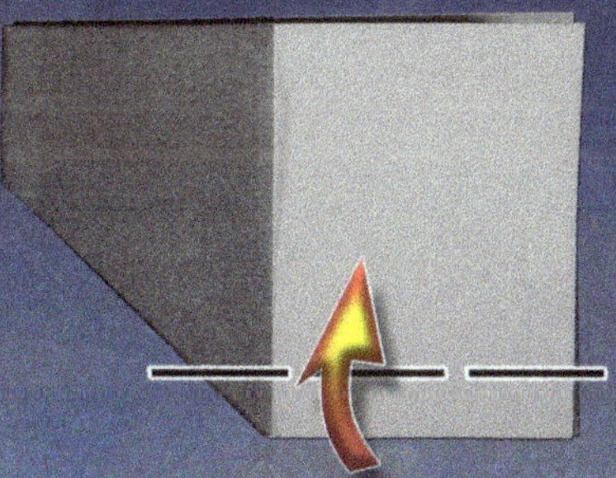

15

Make the other winglet match.

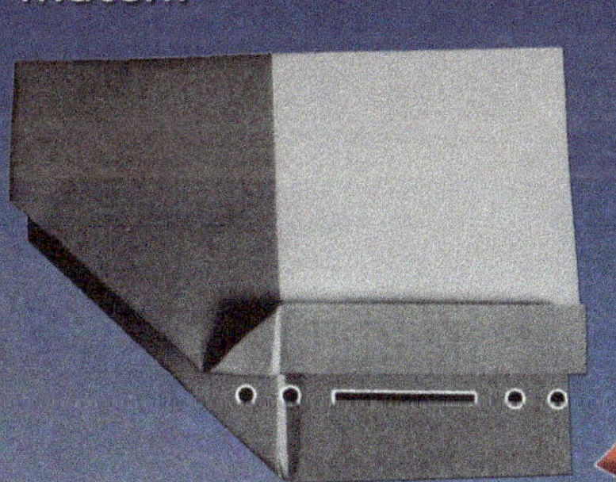

16

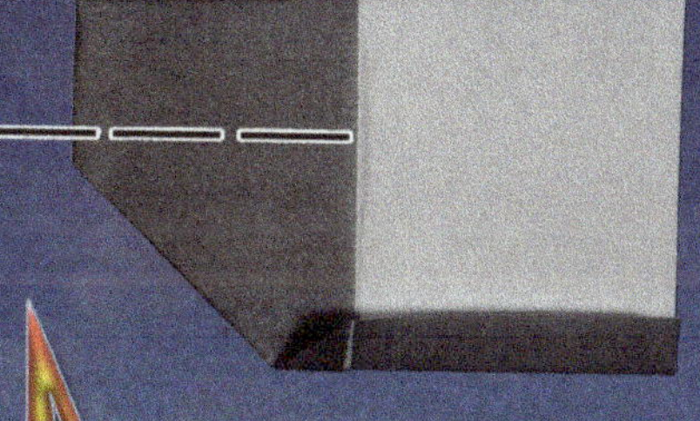

Make a crease that only runs through the layered part and is even with the center crease. Do both wings.

Phat Glider Steps

Add up elevator to the rear of the plane, where the winglet meets the wing.

Take care that the layering matches on both wings. It's okay to sweep out the slack toward the wing tips. You may need to re-make the winglets after you do that.

The wings need to droop a little (negative dihedral angle or anhedral).

Most planes benefit from positive dihedral, which makes them self-correcting. The plane rocks back to neutral when it gets rocked to one side. That's good for a distance or duration plane, but not so good for a stunt plane.

Adjust the wing droop until the plane stays in the banked attitude of your throw.

TUMBLING WING

Magnus Effect (lift generated from a spinning ball or cylinder)? Some think so, but I've always thought of this as simply tumbling, or falling with style.

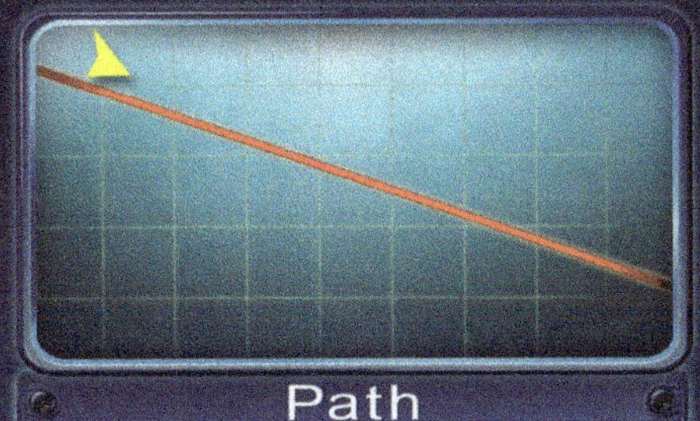

All good Follow Foils move slowly downward (low sink rate) and slowly forward (low wing loading).

I use an 11 x 17 inch piece of cardboard to fly this plane. Some people prefer to learn with larger pieces of cardboard. Ask your doctor if larger cardboard is right for you.

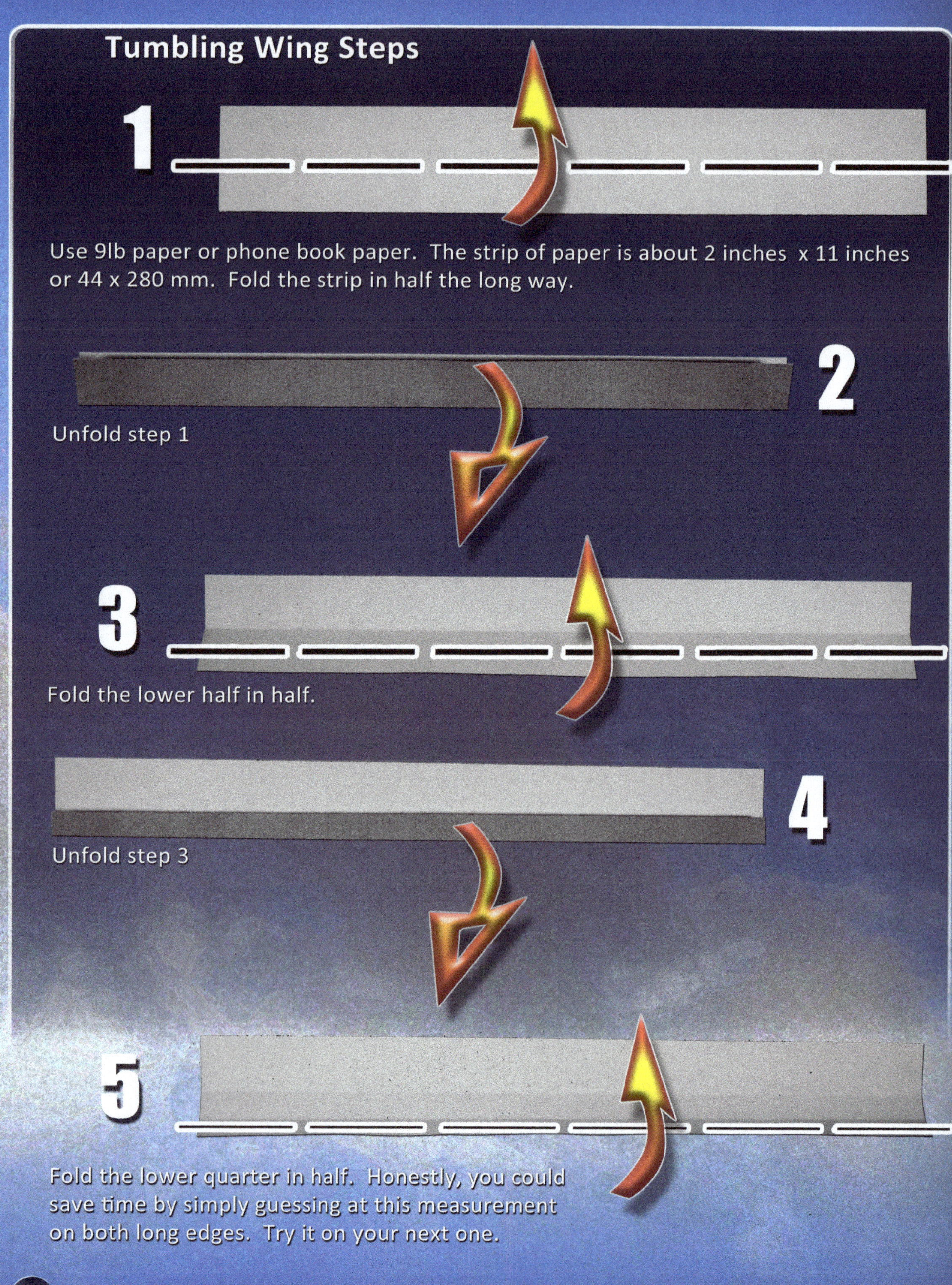

Tumbling Wing Steps

1

Use 9lb paper or phone book paper. The strip of paper is about 2 inches x 11 inches or 44 x 280 mm. Fold the strip in half the long way.

2

Unfold step 1

3

Fold the lower half in half.

4

Unfold step 3

5

Fold the lower quarter in half. Honestly, you could save time by simply guessing at this measurement on both long edges. Try it on your next one.

Tumbling Wing Steps

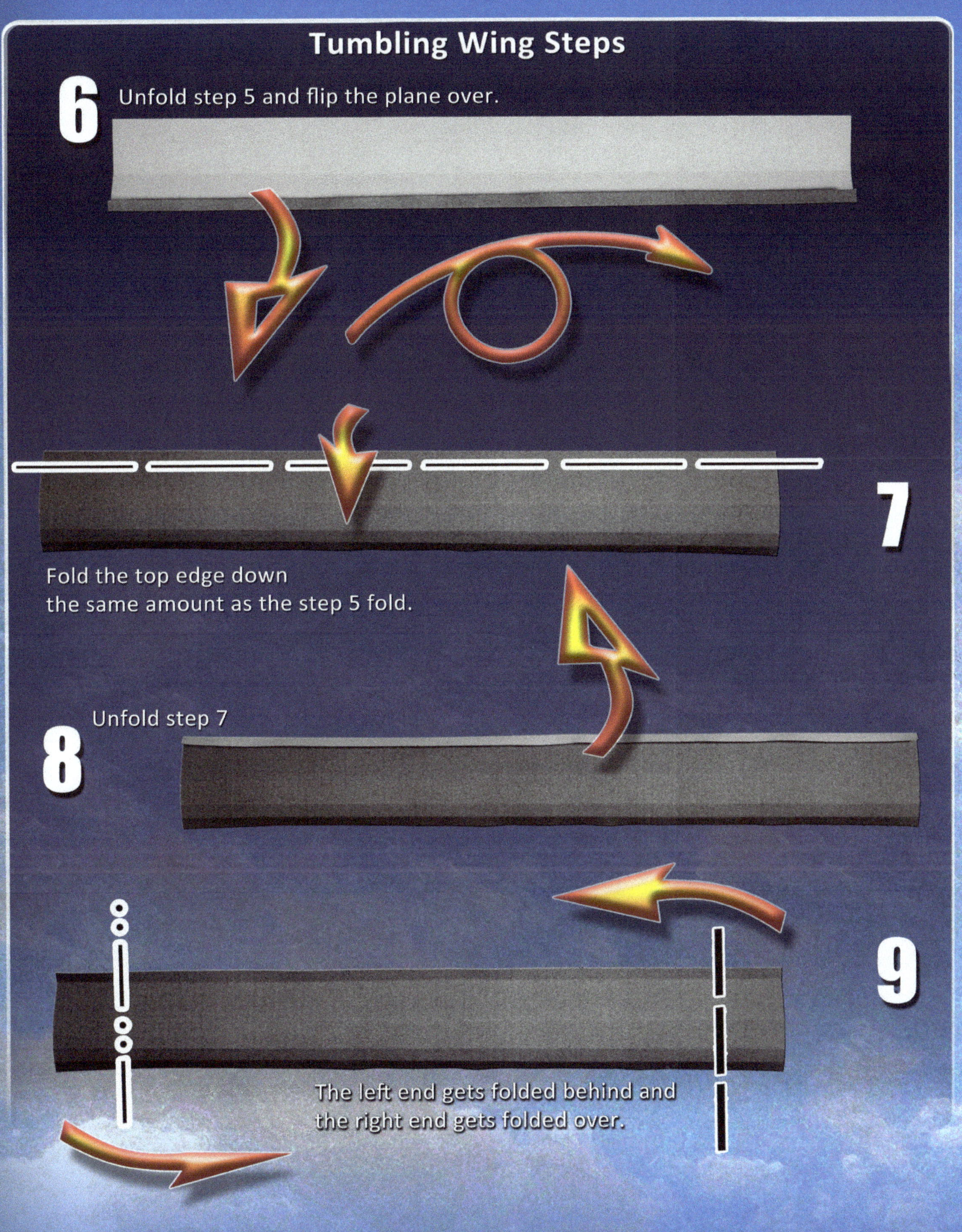

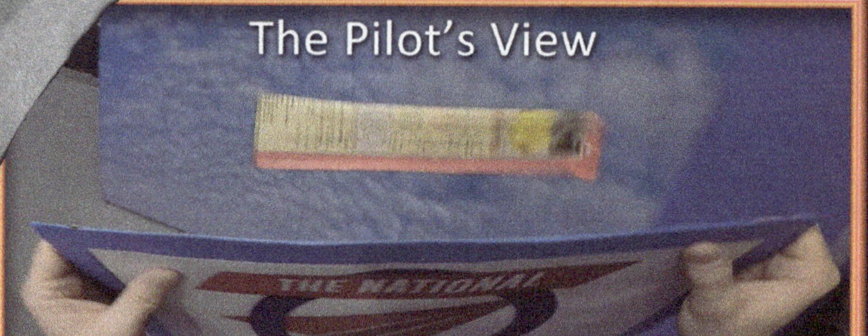

The finished Tumbling Wing should spin backwards as it moves forward.
Hold the wing as shown and give the top a small (emphasis on "small") shove downward to help get it spinning. It should keep tumbling all the way to the floor in a straight line.

If it turns left, add some droop to the right side or flatten out the left side a little.
More droop adds drag. Flattening out a side takes away drag.
When both sides of the plane are moving the same speed, the plane will fly straight.
Next, bring in the cardboard to provide the updraft which balances the sink rate.

Hold the cardboard with the top leaned back. Lean it about half as much (or even less) as this frame from a smoke test. Try to match the forward speed of the Tumbling Wing with your walking speed. As you move foward, the cardboard will scoop air upward. Judging the right speed and tilt for the cardboard takes practice. To turn left, push more air under the right side by pulling the left side of the cardboard back slightly, toward your body.

Walk this way

MAPLE SEED

Take the Maple Seed for a spin

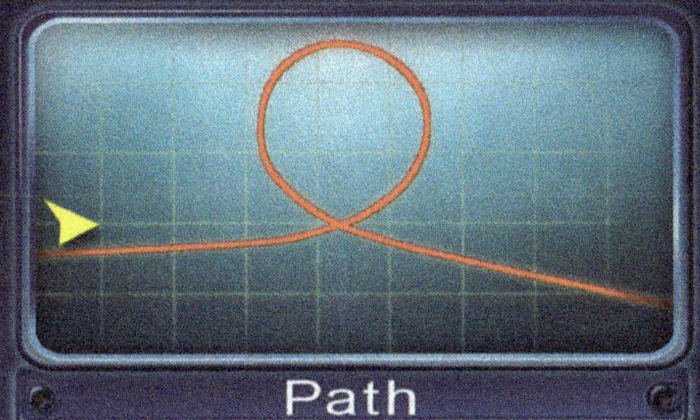

This plane starts out as a knot of paper. Why wait until fall to have fun with a falling object?

Maple Seed Steps

1

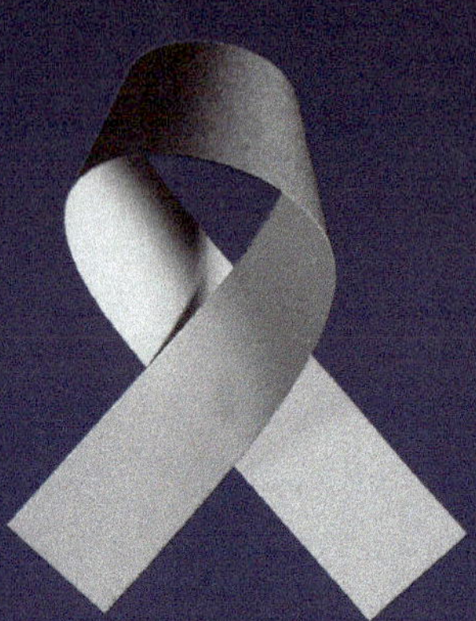

Start with a 25mm wide strip
11 inches long. Give the left
side a half twist clockwise.

2

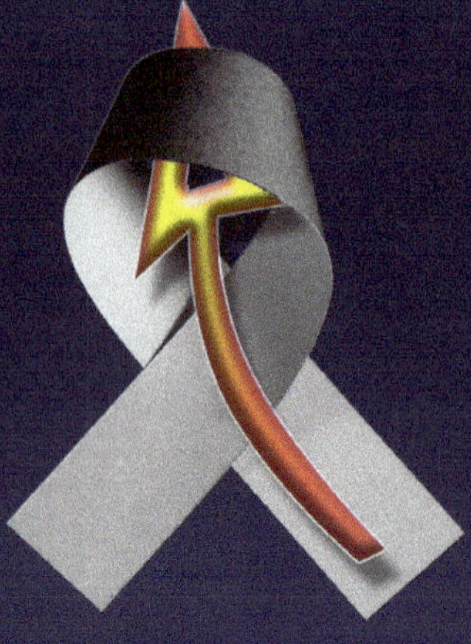

Make a knot by moving the right
end up and through the hole.
Even out both sides before
flattening.

3

Both ends should be the same
size. Rotate the plane so the
center of the knot is up.
Fold the center down to
meet the lower left corner.
the crease goes through
the upper left corner.

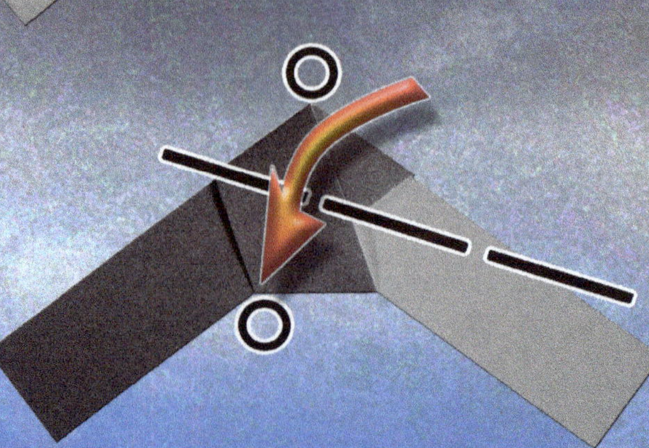

Maple Seed Steps

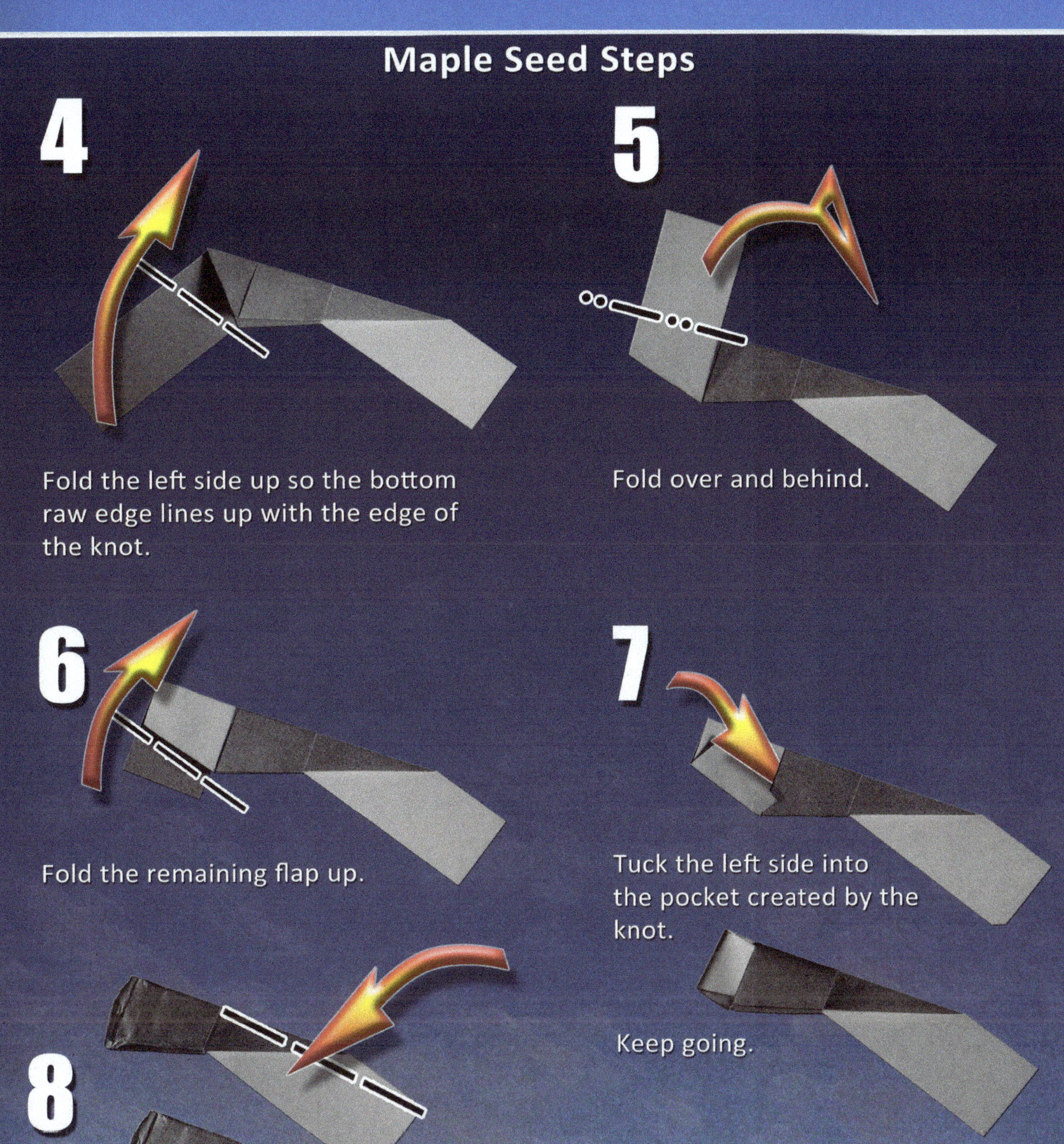

4

Fold the left side up so the bottom raw edge lines up with the edge of the knot.

5

Fold over and behind.

6

Fold the remaining flap up.

7

Tuck the left side into the pocket created by the knot.

Keep going.

8

Fold down from the corner of the knot to the corner of the wing.

RB DART

The winningest plane in Red Bull competition history.

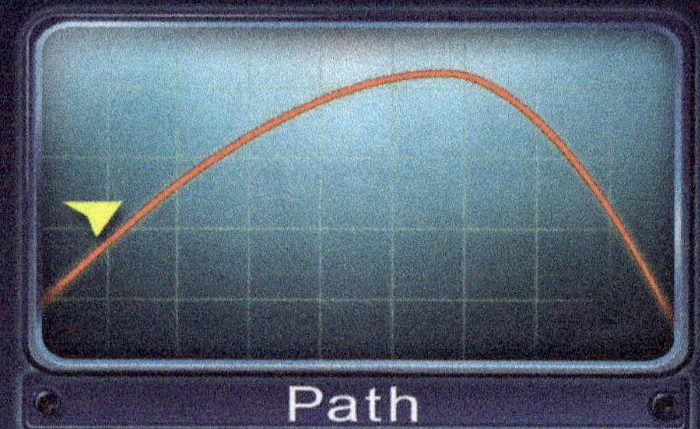

Simple to fold. Brut-force propelled. Unconventional.

RB Dart Steps

1

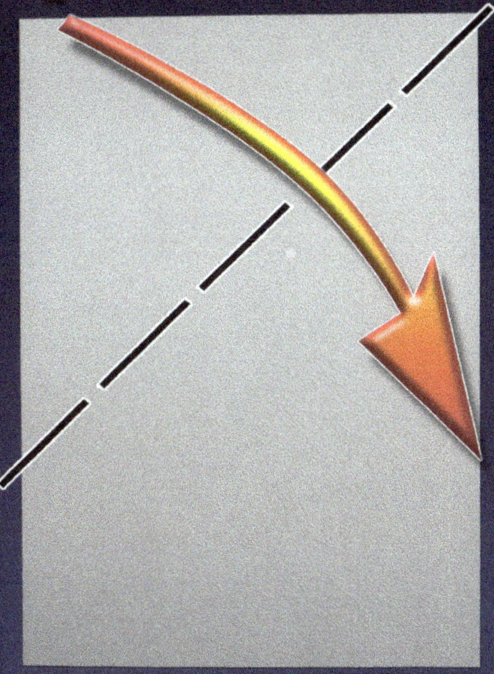

Start with A4 paper.
To convert US letter size to the
right ratio, remove 19mm from
one side of the paper.
Make a diagonal fold by lining
up the top with the left side.

2

Mark the center of the crease.

2A

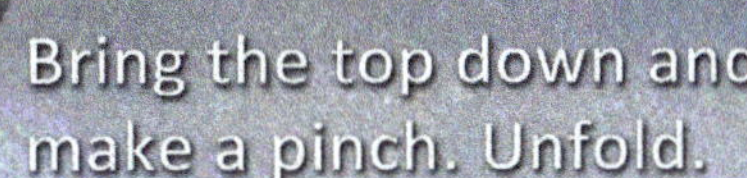

Bring the top down and
make a pinch. Unfold.

3

Fold from the
pinch to the
corner.

4

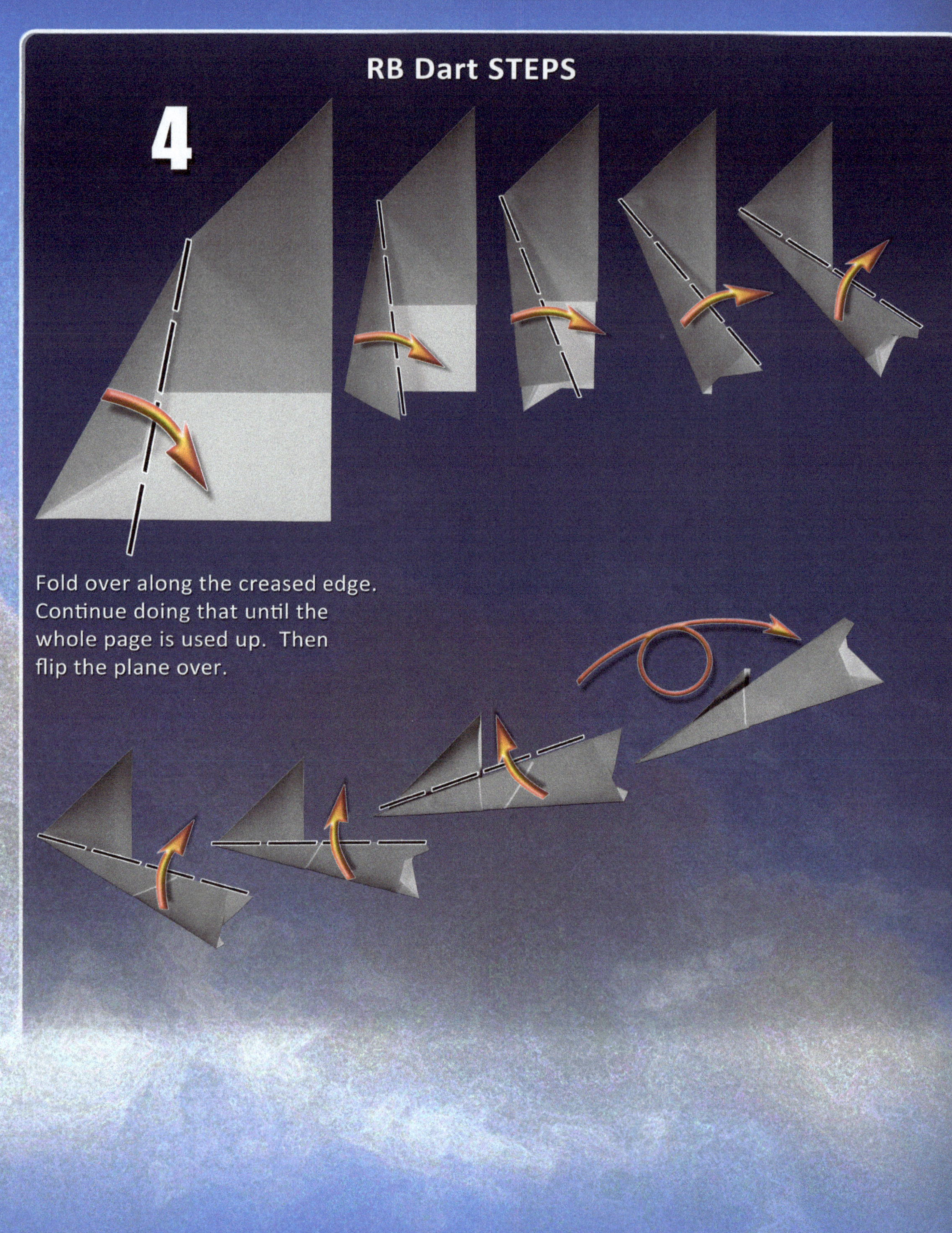

Fold over along the creased edge.
Continue doing that until the
whole page is used up. Then
flip the plane over.

RB Dart Steps

5

Tuck the loose corner under the layers. Easy as one, two, three.

Fold the plane in half to finish.

6

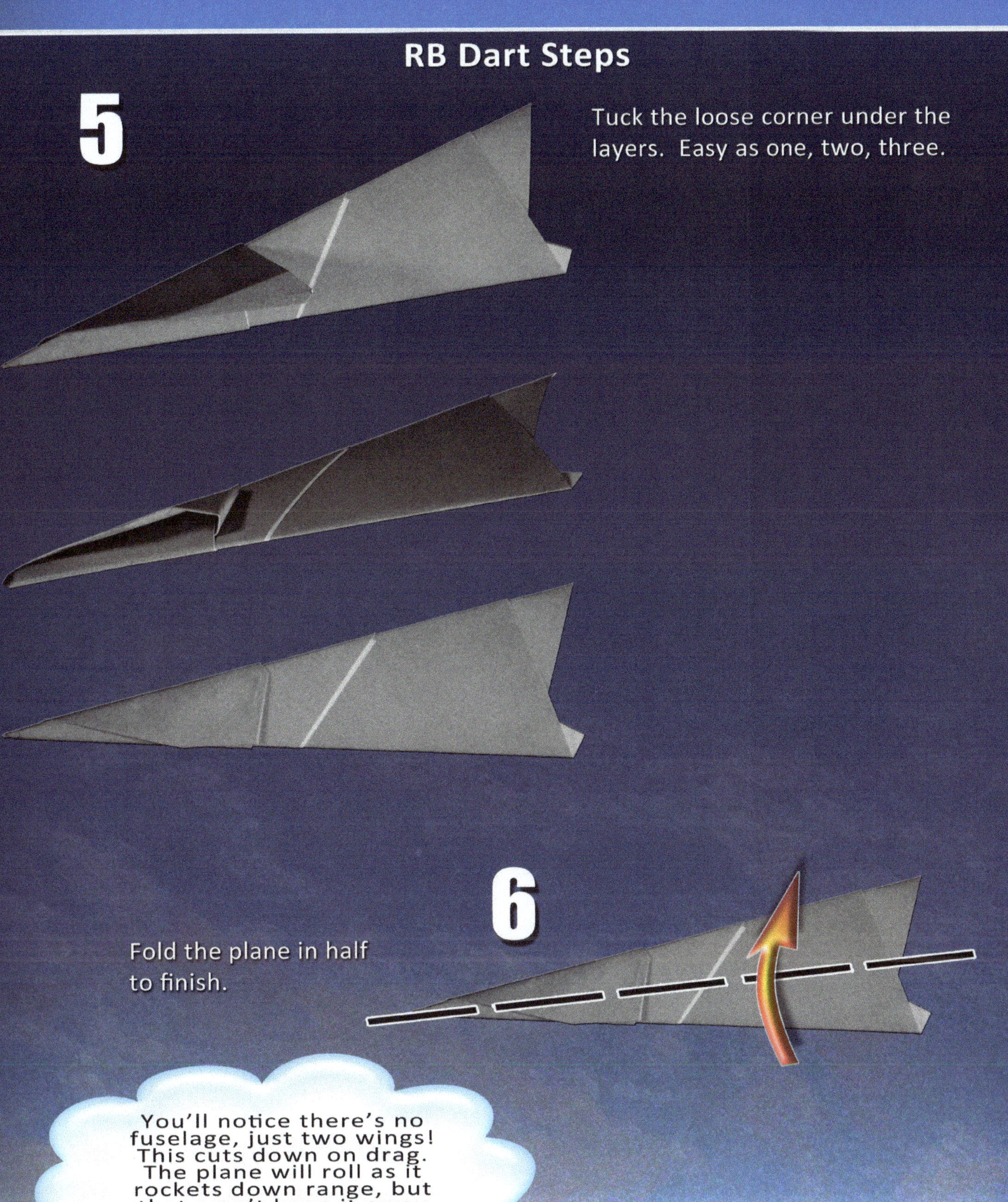

You'll notice there's no fuselage, just two wings! This cuts down on drag. The plane will roll as it rockets down range, but that won't keep it away from the winner's circle.

The Tube

Question: Where on the surface of the plane is the Center of Gravity?

Answer: Nowhere! The CG is at the center of the opening.

The more spin you give it, the straighter it flies. Let it roll off your fingertips. Put some strength into the throw. I use this airplane to talk about Center of Gravity because the CG isn't on the plane; it's in thin air! Magnus Effect, Precessional Forces, and Centrifugal Force are all possible to explore.

The Tube Steps

1

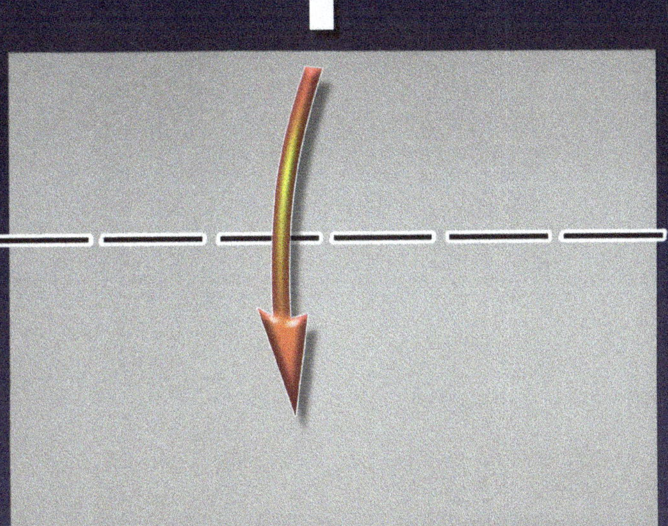

Start with a long side up. Fold 1/3 down. That means the layered part will end up the same size as the unlayered part. It's okay for layered part to be just a little bigger than the unlayered part

2

Fold the layered part in half.

3

Again, fold the layered part in half.

4

Rub the layered part over edge of a table to bend the thick layers into a curve. Then unfold step 3

The Tube Steps

5

Put the left side inside the right end. The loop of paper stays in front, the two corners go behind. If you're left handed, put the right side inside the left end. The overlap should be 1" to 1.5".

6 Remake the step 3 crease. It will get ugly for a moment before it looks nice.

7 Fold about .25 inch (.5 cm) of the rear to the inside to lock it together.

The Swan

A slow, graceful glider; The Swan is part origami and part aircraft. The word aero-gami comes to mind.

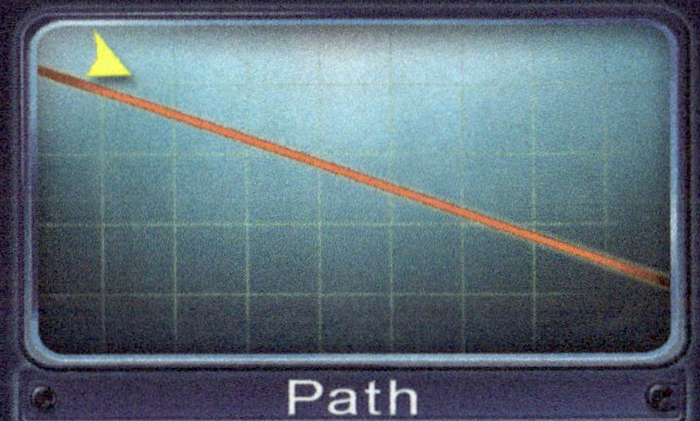

You may need a small bit of up elevator. Use a gentle throw.

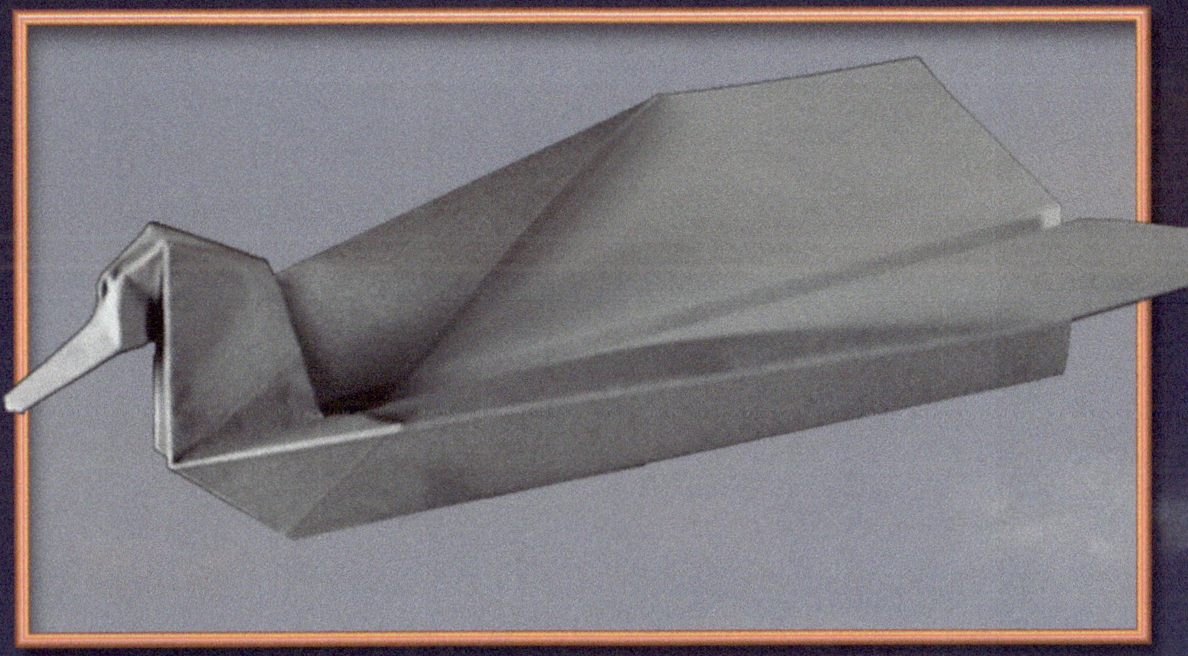

The Swan Steps

1

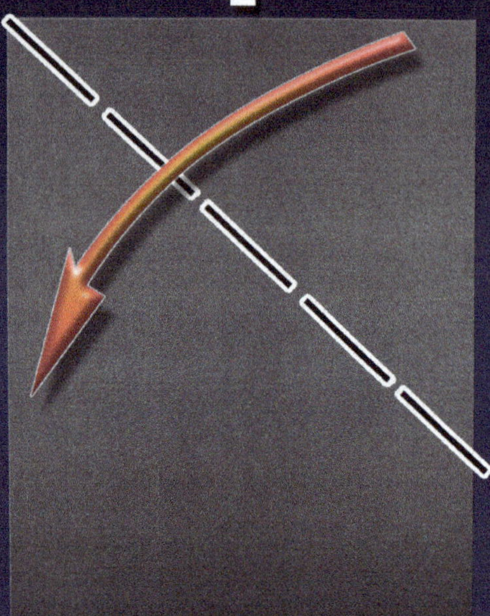

Make a diagonal fold. Line up the top with the left side.

2

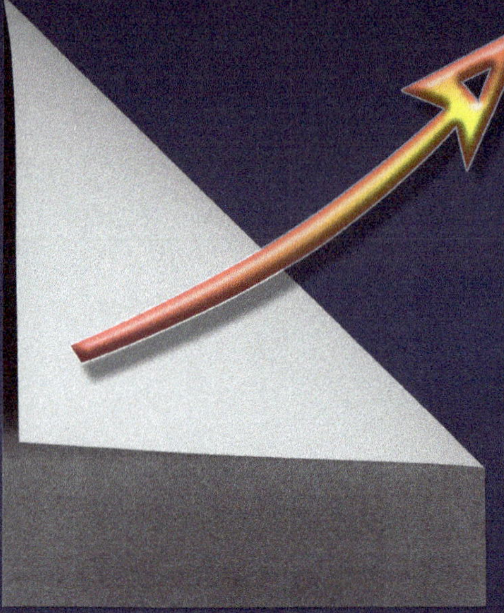

Unfold step 1

3

Fold and unfold the other diagonal.

4

Flip the page over.

The Swan Steps

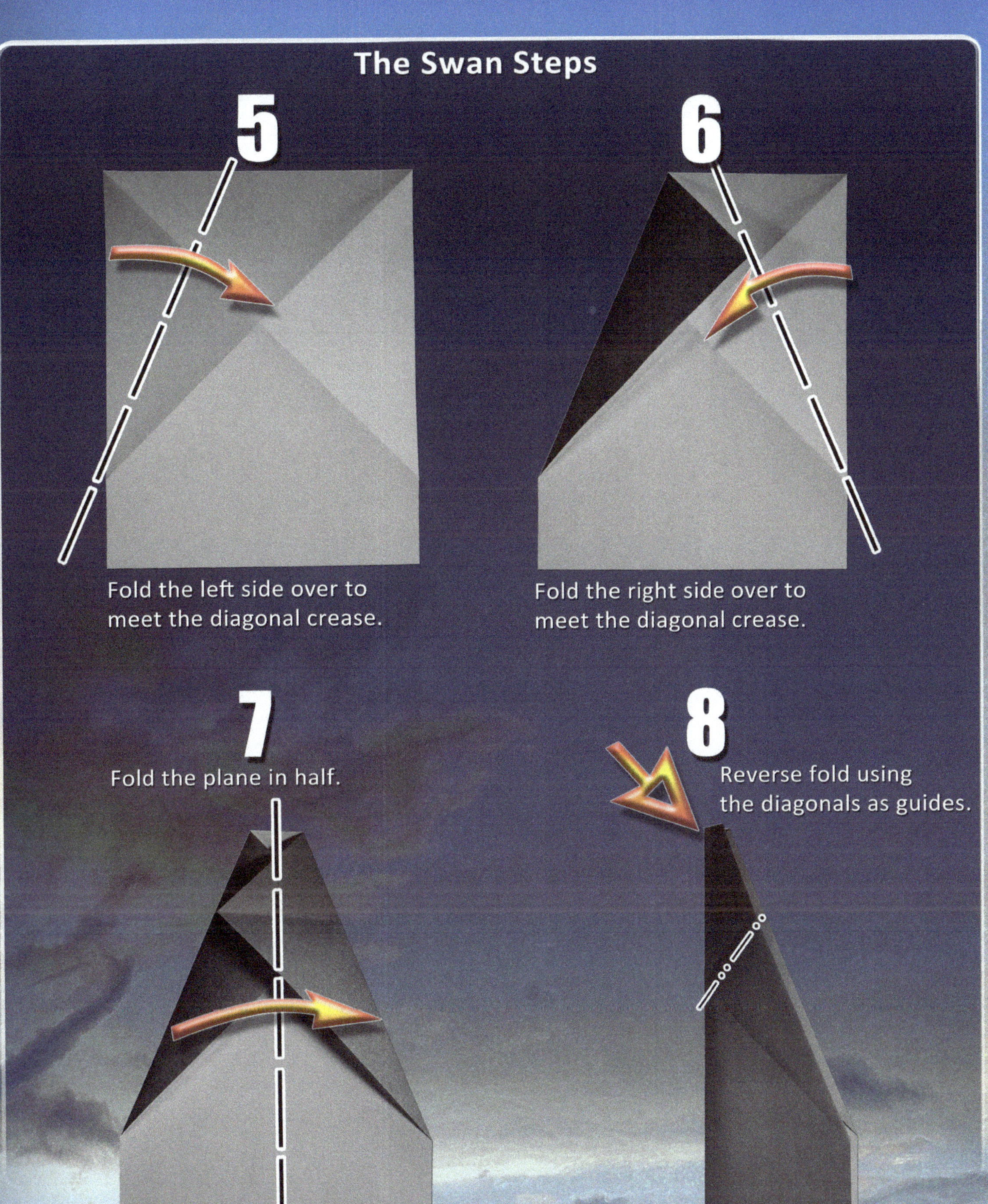

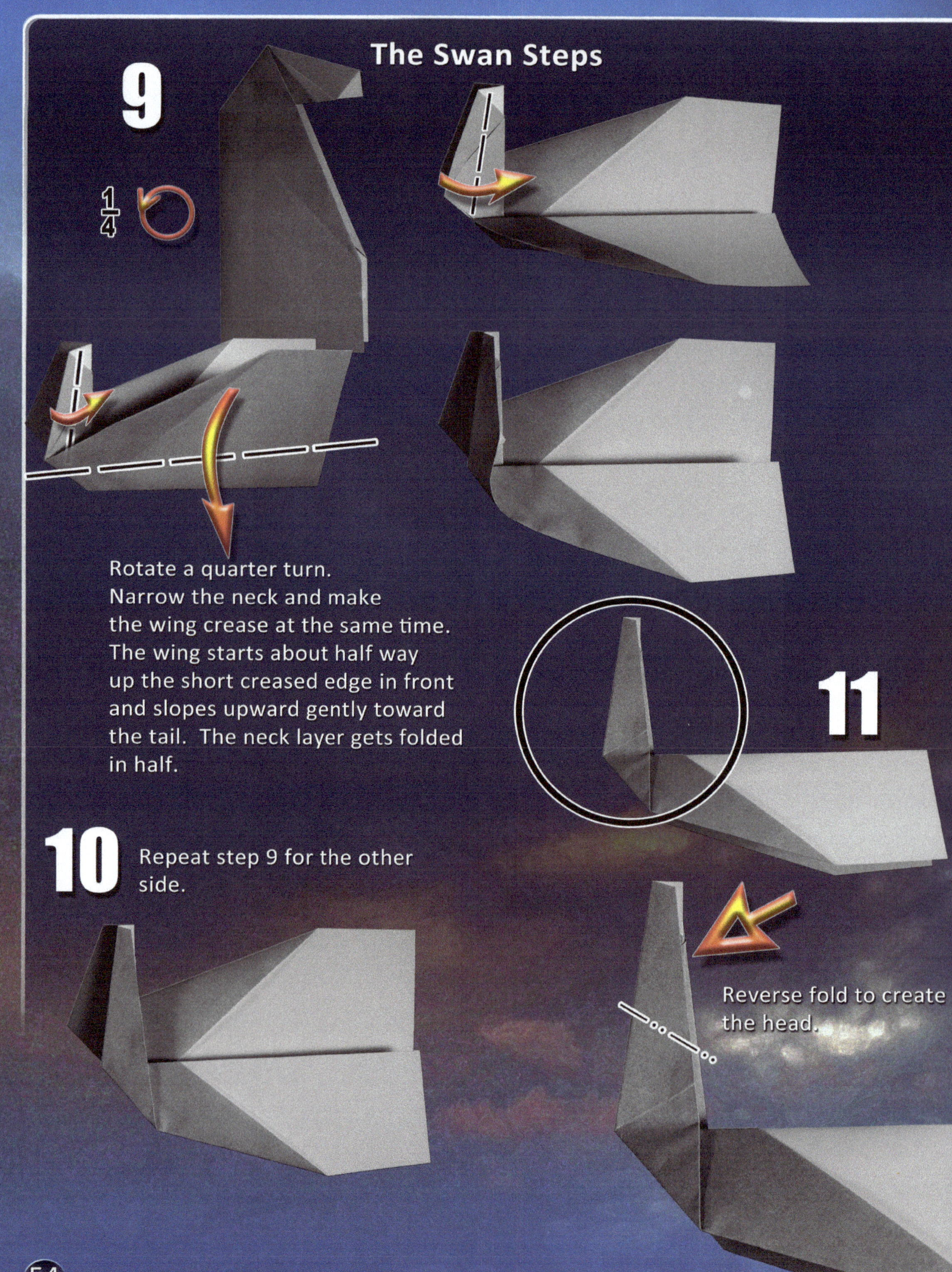

The Swan Steps
9
1/4
Rotate a quarter turn.
Narrow the neck and make
the wing crease at the same time.
The wing starts about half way
up the short creased edge in front
and slopes upward gently toward
the tail. The neck layer gets folded
in half.
11
10
Repeat step 9 for the other
side.
Reverse fold to create
the head.

The Swan Steps

12

Pinch the bill flat so that the layers flatten outward on both sides.

The middle picture is the result.

Get fancy by rounding off the bill. Fold the corners under like the bottom photo.

Lifting the wings will make the creases where the neck meets the wings.

Have a naming contest. I'll start: Dick Van Duck. See? You can certaininly beat that. Go ahead. Make the plane and challenge your friends.

Bat Plane

The Bat Plane flaps its wings as it flies. That's enough to make it worth learning.

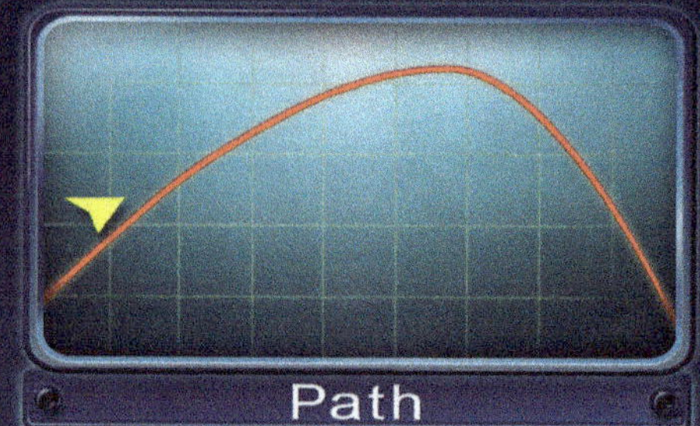

There's a move near the start that collapses the layers together. That's the hardest part of the folding. The adjusting takes patience.

Bat Plane Steps

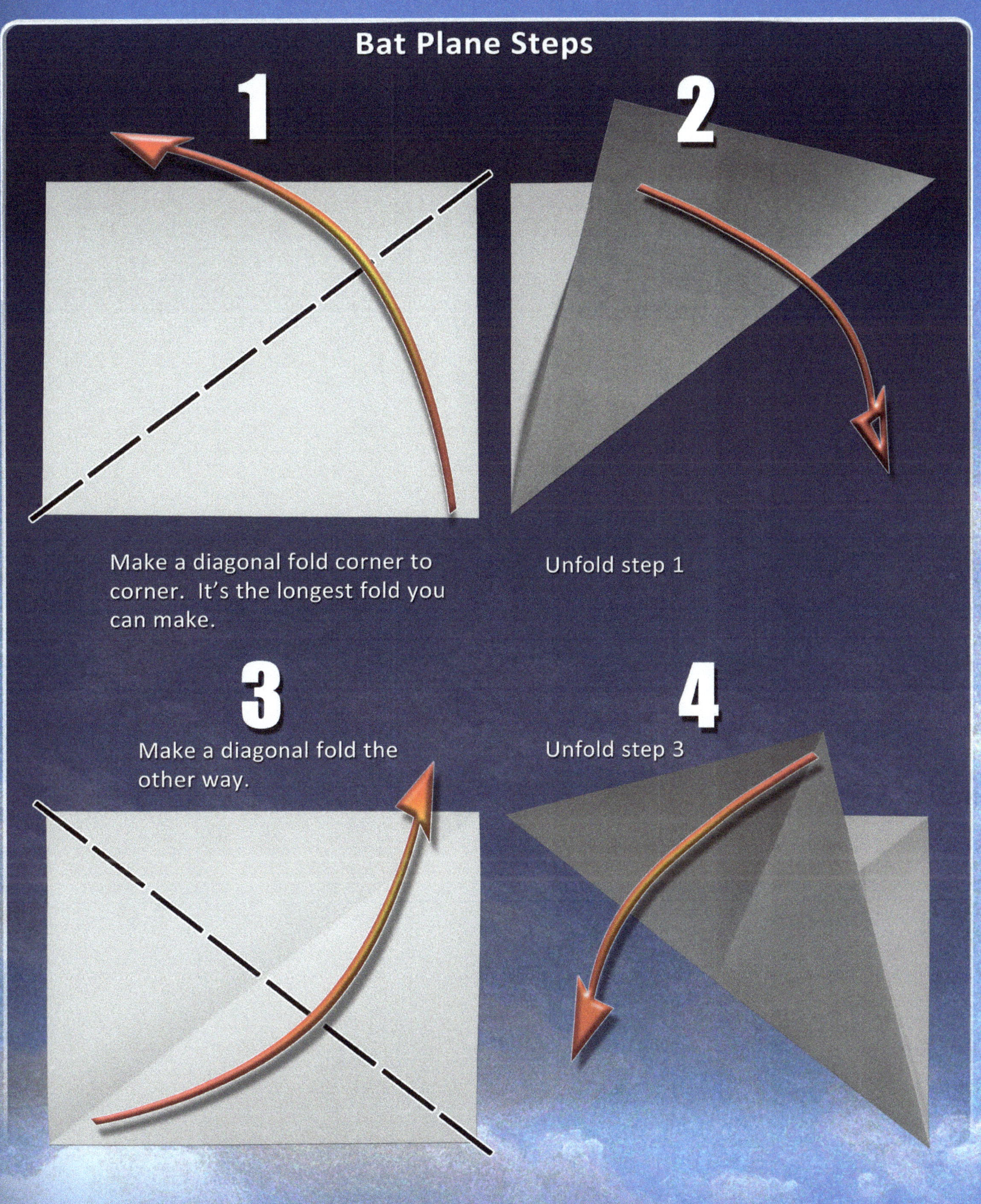

1

Make a diagonal fold corner to corner. It's the longest fold you can make.

2

Unfold step 1

3

Make a diagonal fold the other way.

4

Unfold step 3

Bat Plane Steps

5

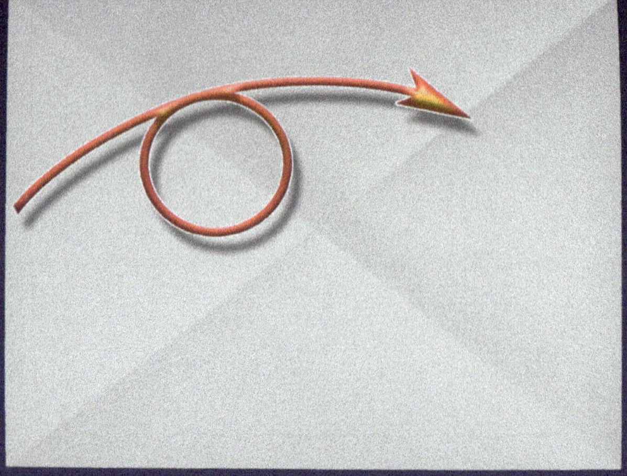

Flip the page over.

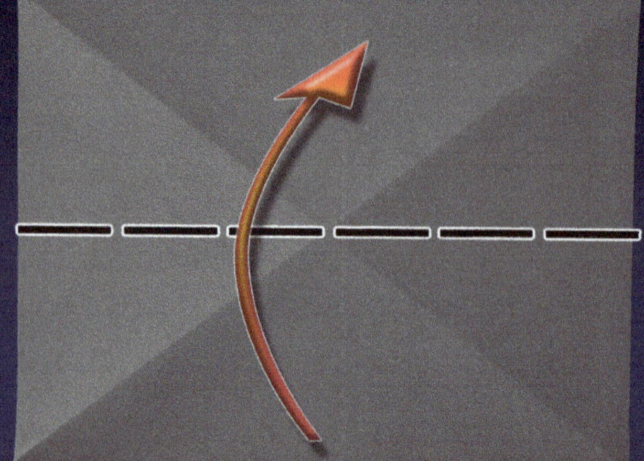
6

Fold the page in half.

7

Unfold step 6, and stand up the page making a long tunnel.

8

Press down where all the creases meet, until the center "pops".

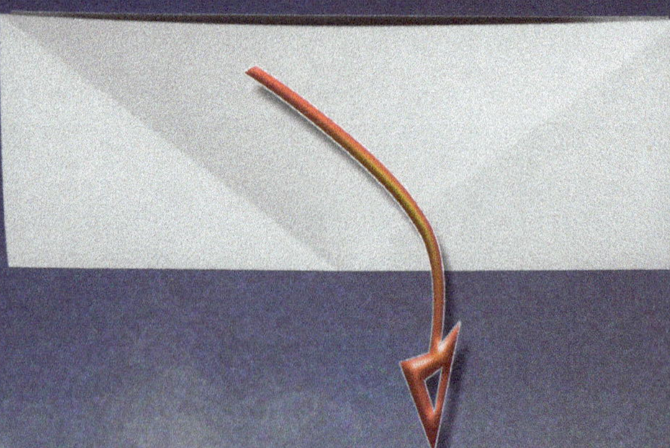

9

Bring the sides in and the top down.

10

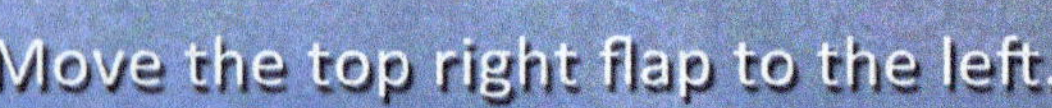

Move the top right flap to the left.

Bat Plane Steps

11

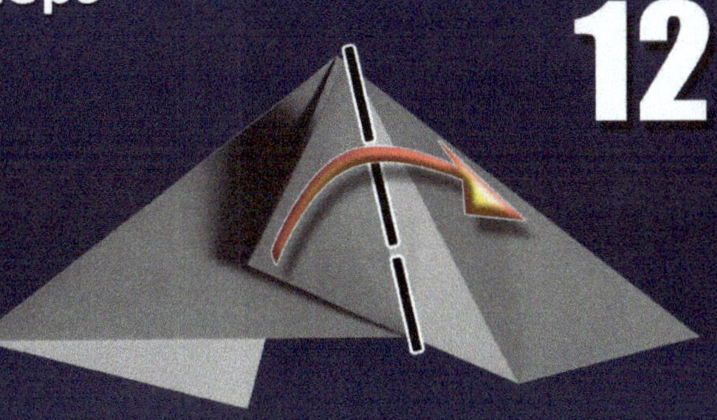

Fold the point to the top.

12

Fold along where the layer ends.

13 10,11,12 repeat for this side starting with moving the flap to the right.

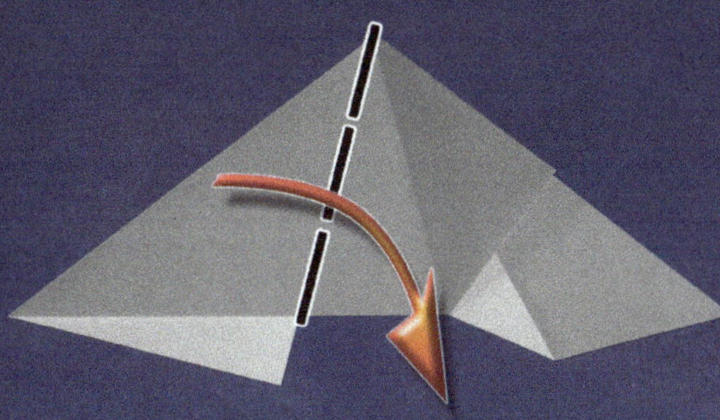

14

Fold the corner to the top.

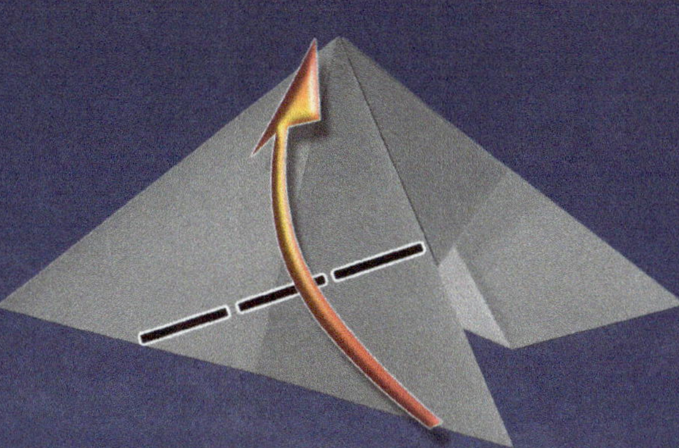

15 Move the corner to the left using the end of the layer as a guide.

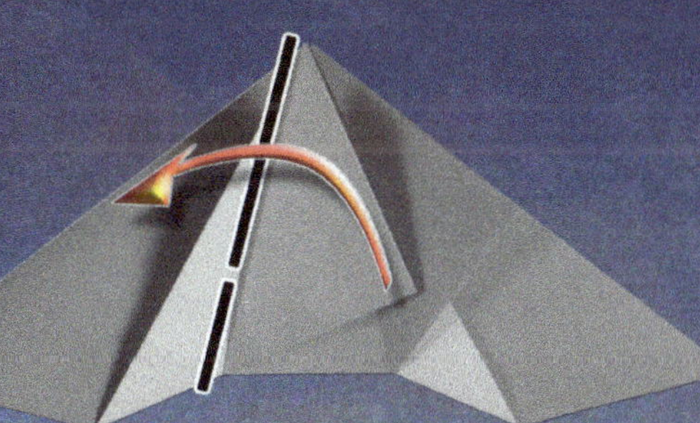

Fold down across the two corners.

16

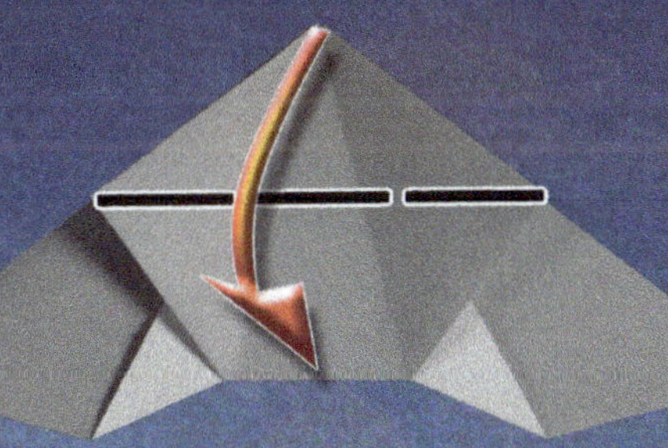

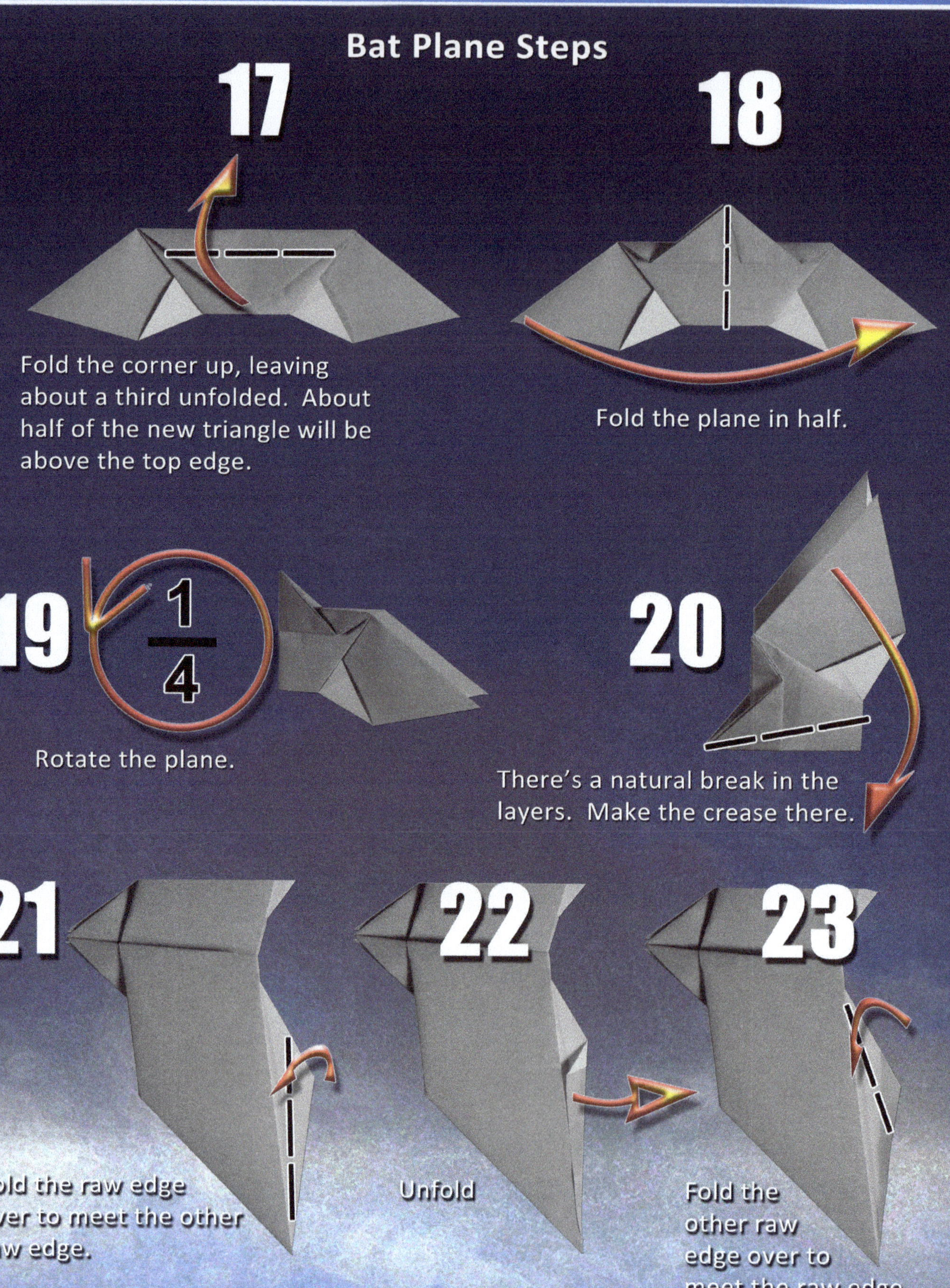

Bat Plane Steps

17

Fold the corner up, leaving about a third unfolded. About half of the new triangle will be above the top edge.

18

Fold the plane in half.

19

1/4

Rotate the plane.

20

There's a natural break in the layers. Make the crease there.

21

Fold the raw edge over to meet the other raw edge.

22

Unfold

23

Fold the other raw edge over to meet the raw edge.

Bat Plane Steps

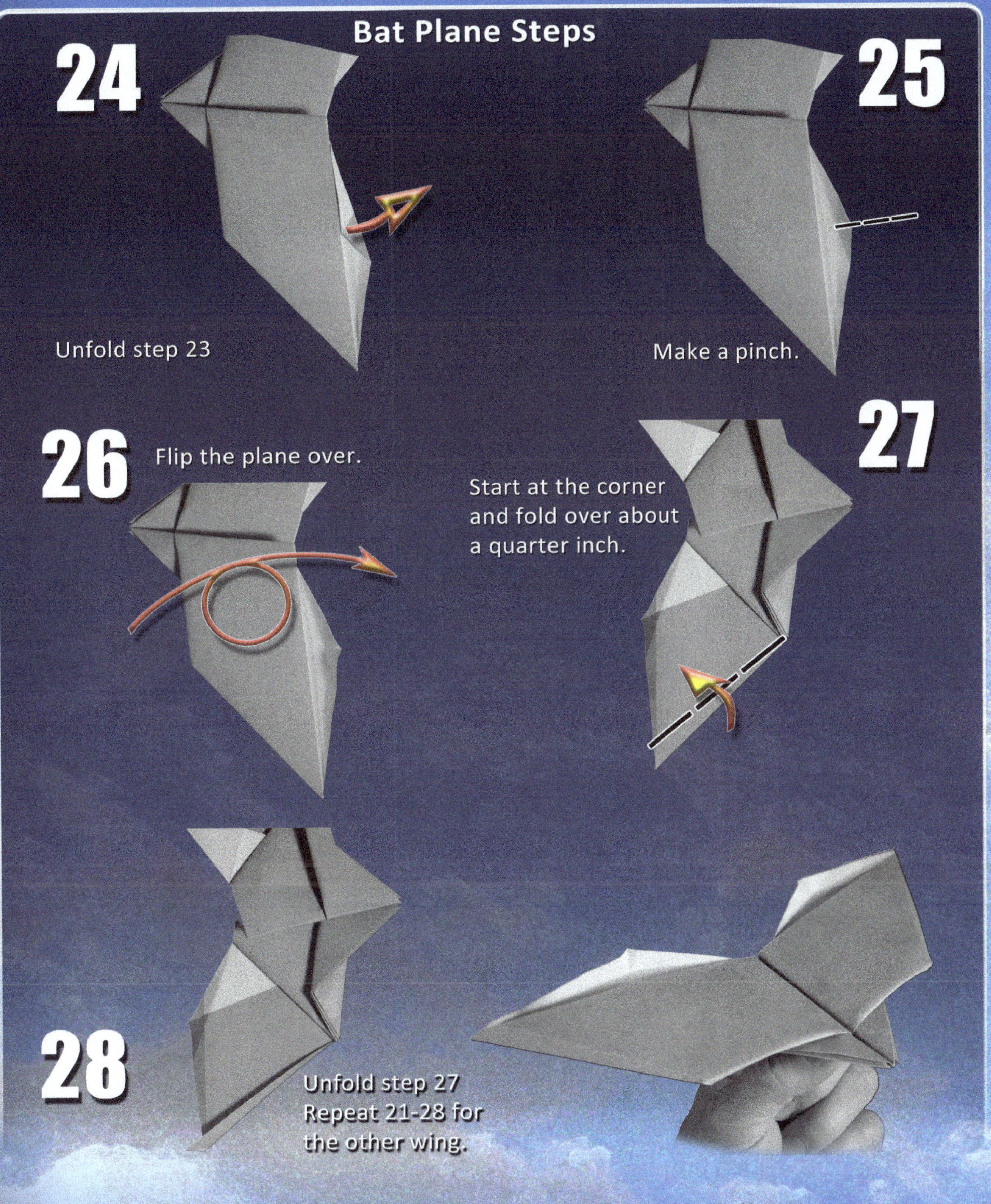

24 Unfold step 23

25 Make a pinch.

26 Flip the plane over.

27 Start at the corner and fold over about a quarter inch.

28 Unfold step 27 Repeat 21-28 for the other wing.

ULTRA GLIDE

Glider

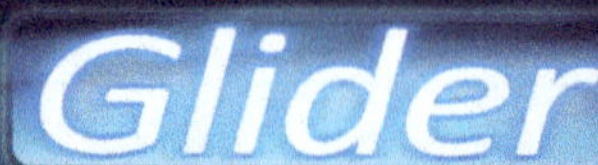

Flawless, smooth glides with the efficiency of a locked together fuselage.

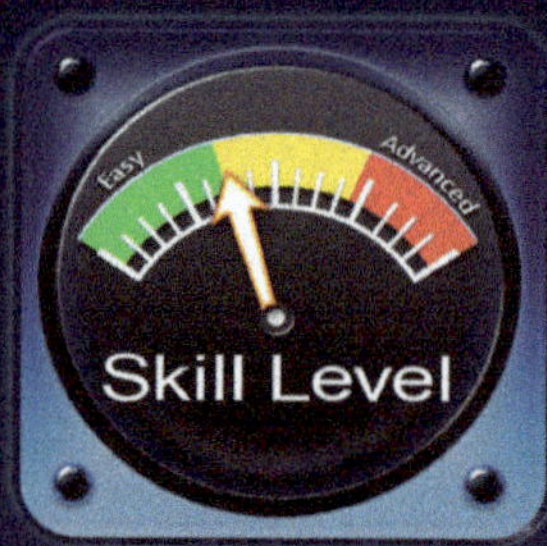

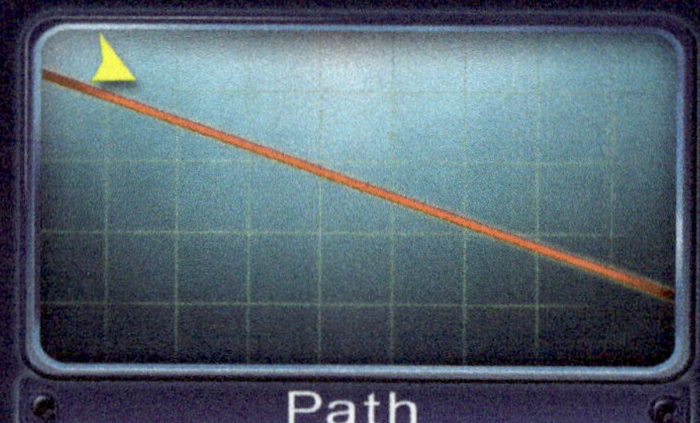

A little finesse is needed for the nose locking move, and the rest is simple paper airplane making.

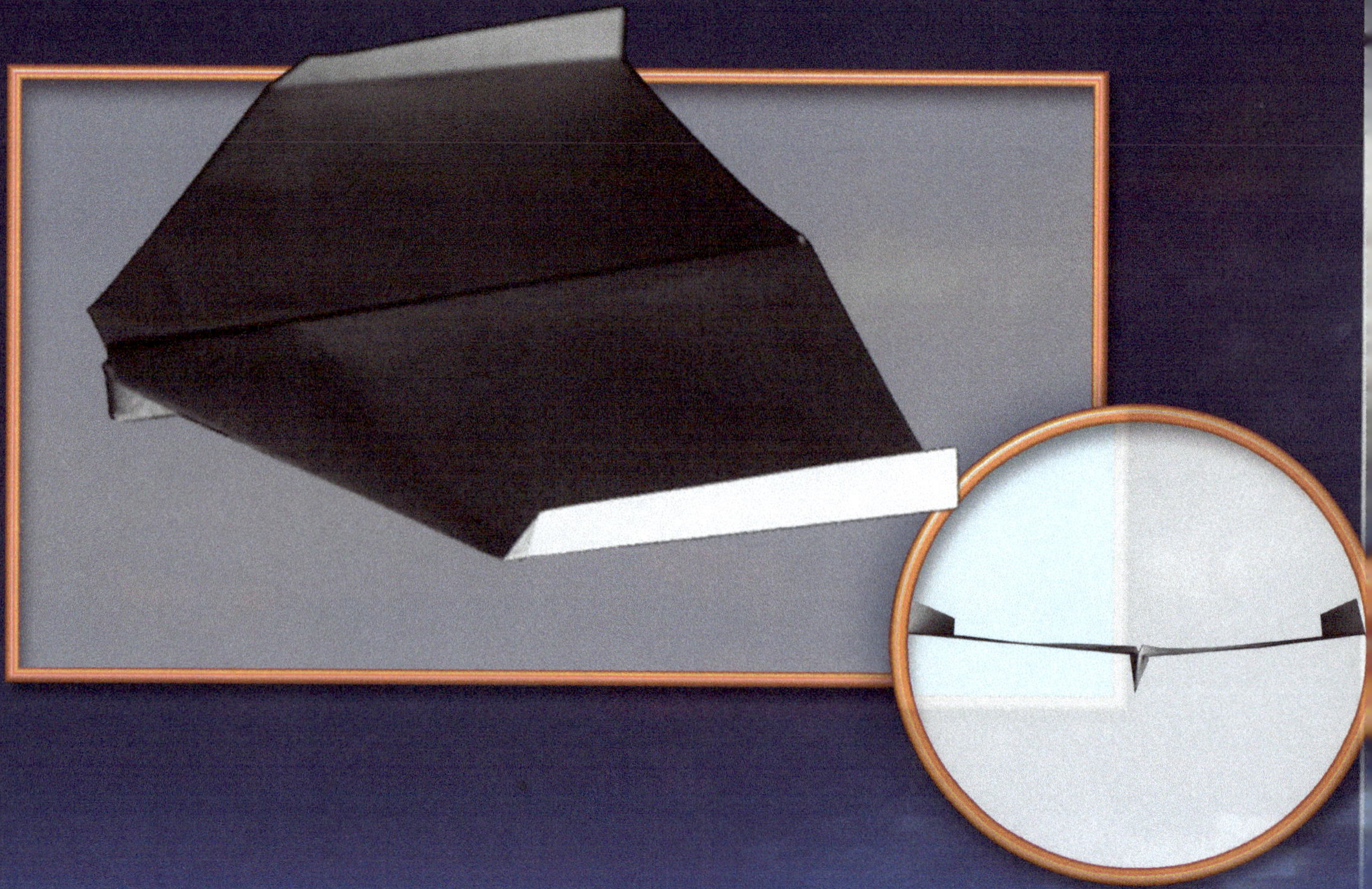

Ultra Glide Steps

1

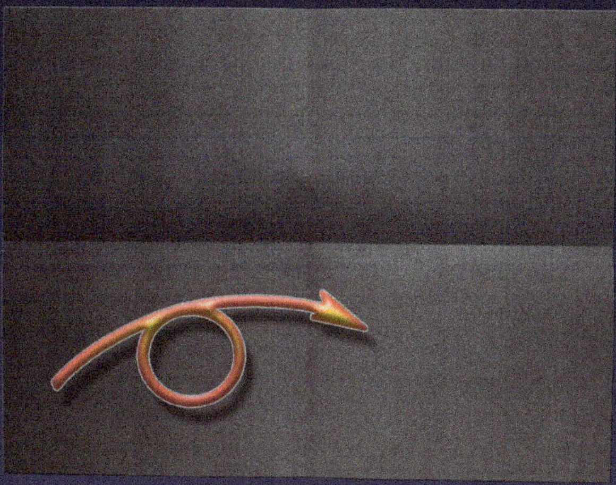

Start with a long side on top. Fold in half; top to bottom and left to right. Then flip the page over.

2

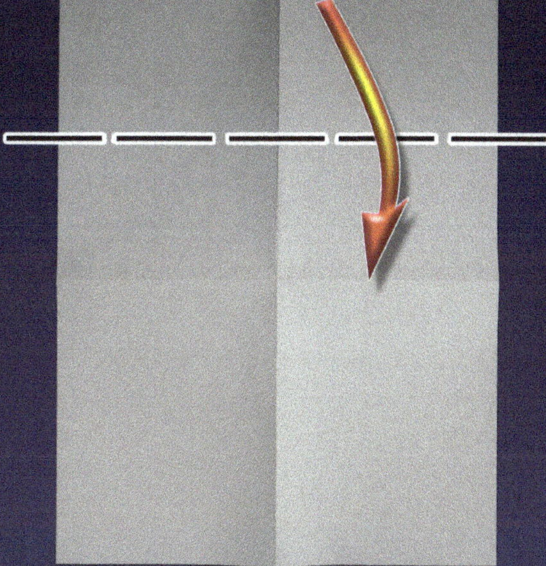

The short side is now the top. Fold the top down to meet the center crease.

3

Fold the top edge down to the center.

4

Unfold step 3

Ultra Glide Steps

5

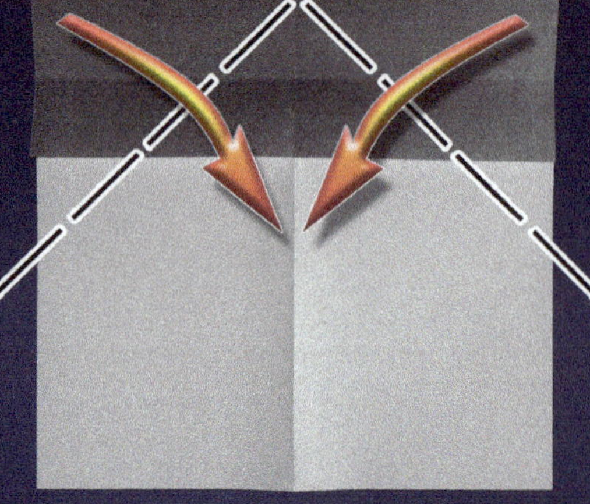

Fold the top corners to the center.

6

Unfold step 5

7

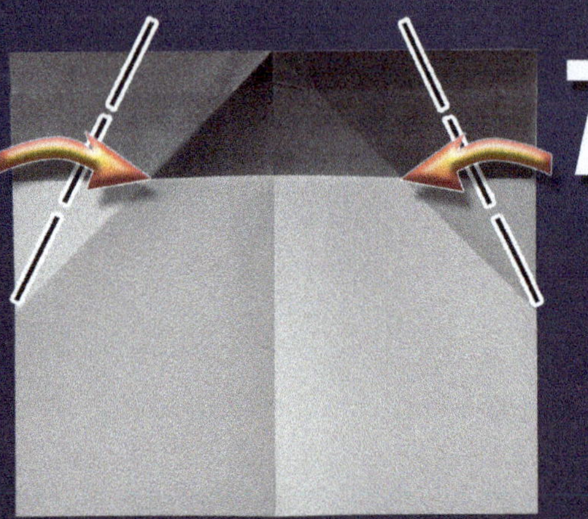

Fold the outside edges to the creases.

8

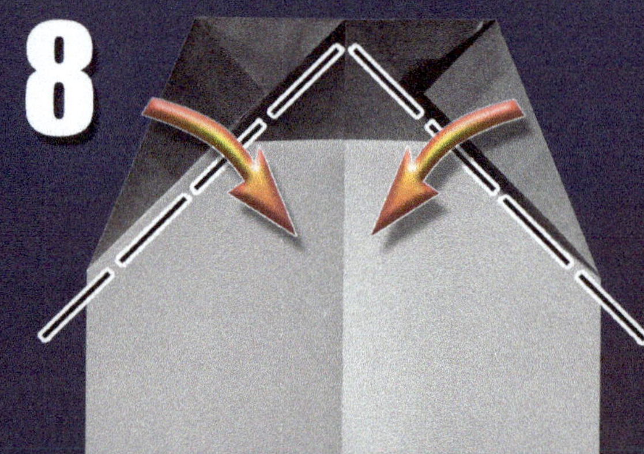

Remake the step 5 creases.

9

Fold the top down to meet the two corners.

10

Unfold step 9

11

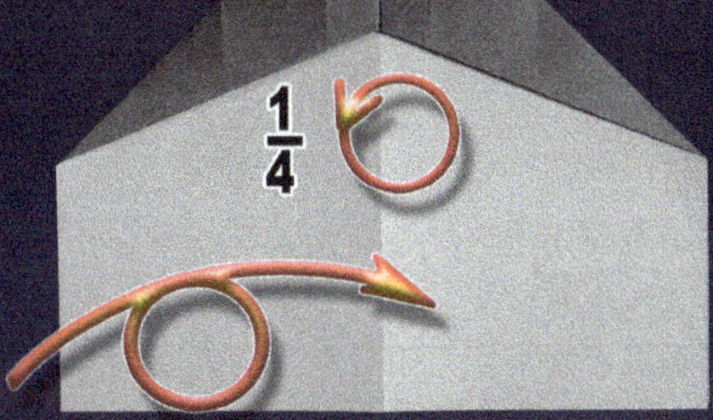

Flip the plane over and rotate the nose to the left.

12

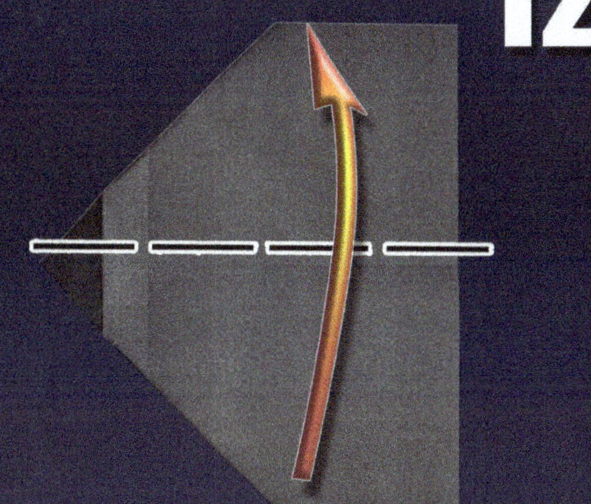

Fold in half, taking care to line up the rear corners.

13

Use the crease from step 10 to make a squash fold. The image in front is zoomed in.

14

Fold the top layer of the squash in half by folding it down.

15

Fold the left half of the squash behind.

16

The wing crease.

Start at the top of the squash, and the crease continues to the rear corner.

17

Make the other wing match.

18

Make the winglet parallel with the wing crease.

19

Make the other winglet match.

Adjusting Ultra Glide:

Use a flat or very slight upward sweep on the wings. Add up elevator at the center of the tail, where the wings meet. Use gentle throws until you get a straight, flat glide.

MAX LOCK

A new nose locking technique and big wingspan make this a fun plane to fold and fly.

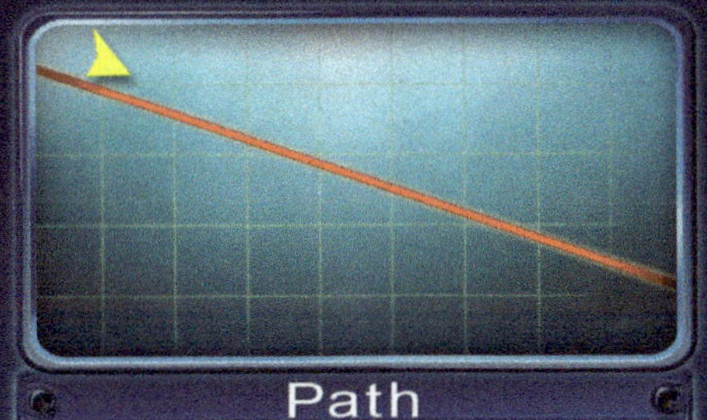

A bit trickier than Ultra Glide and even smoother through the air, if that's possible.

1

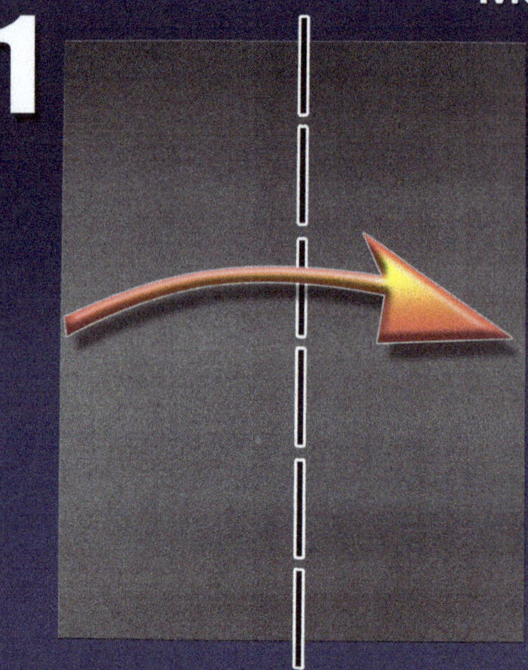

Start with a short side up.   Fold the page in half.

2

Unfold step 1

3

Fold the left edge to the center.

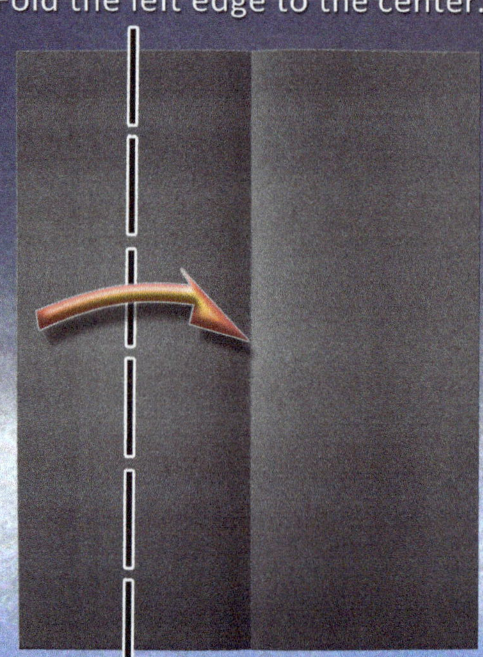

4

Unfold step 3

Max Lock Steps

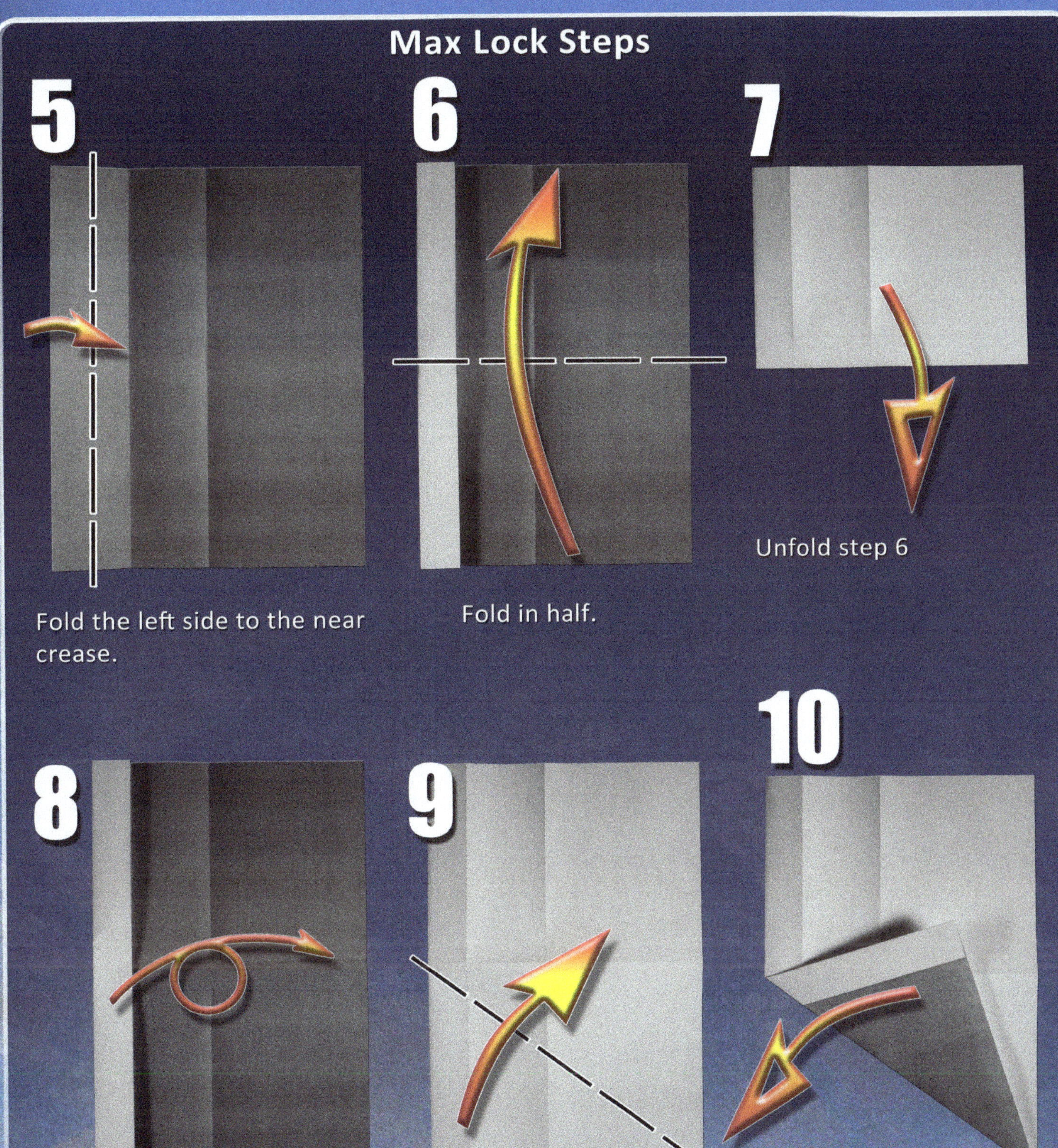

5

Fold the left side to the near crease.

6

Fold in half.

7

Unfold step 6

8

Flip the plane over.

9

Fold from the center to the corner.

10

Unfold step 9

Max Lock Steps

11

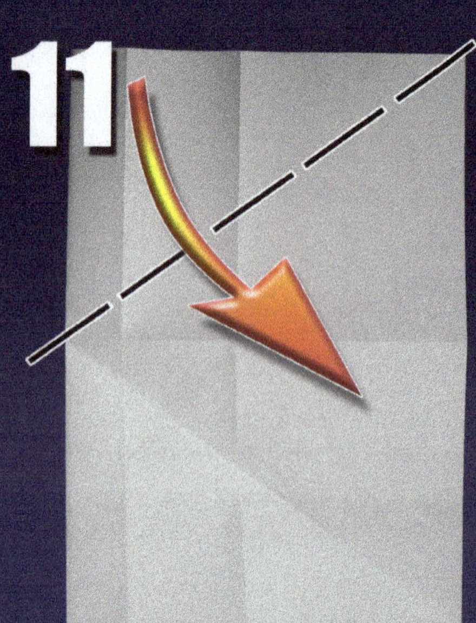

Fold from the center to the corner.

12

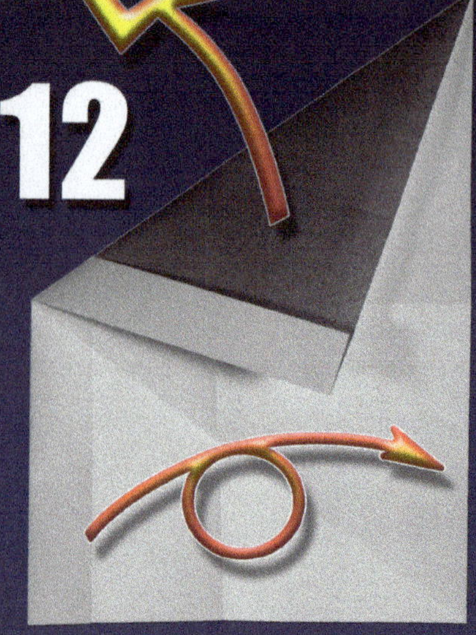

Unfold step 11 and flip the plane over.

13

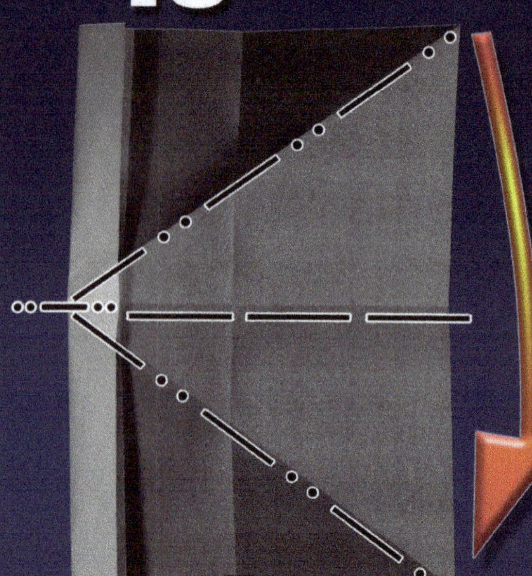

This is a giant reverse fold. The top layer on the left comes upward. The bottom layer moves down as the diagonals move upward.

14

Pull out and down on the thin layer, while pushing the small crease in line with the longer creased edges.

Notice how the top layers from step 14 got rolled outward to pull the front crease down. This is the nose lock. Now lift the whole wing upward.

15

Max Lock Steps

16

Move the top edge to the crease.

17

Fold the front edge
to the crease.

18

Remake the crease.

19

Flip the plane over
and repeat 15-18

20

The wing crease goes from the top of the lock
to the corner of the tail.

Max Lock Steps

21

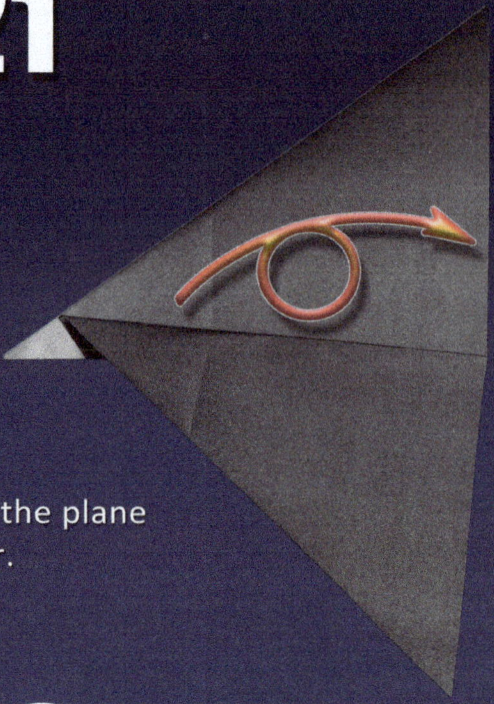

Flip the plane over.

22

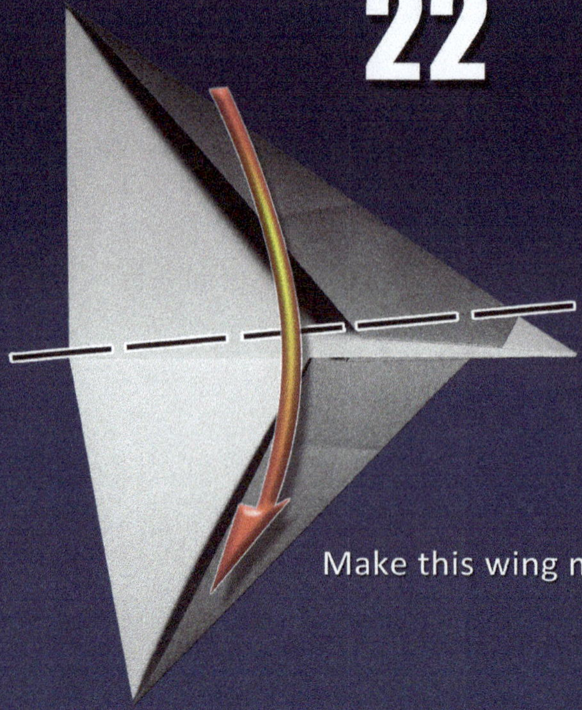

Make this wing match.

23

Fold up about a third of the wing for the winglet.

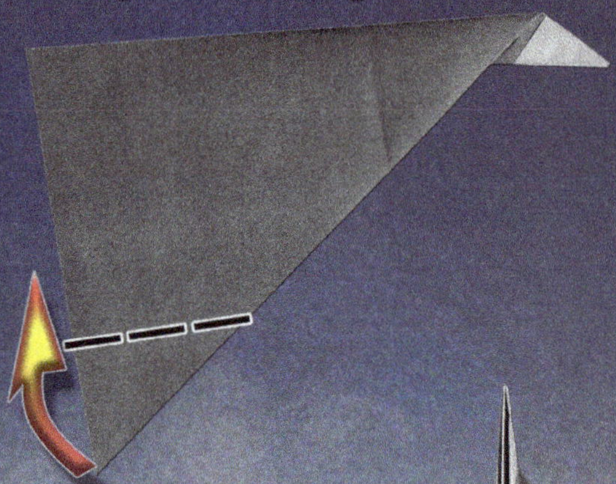

24

Make the other winglet match before opening up the wings and placing the plane on its back.

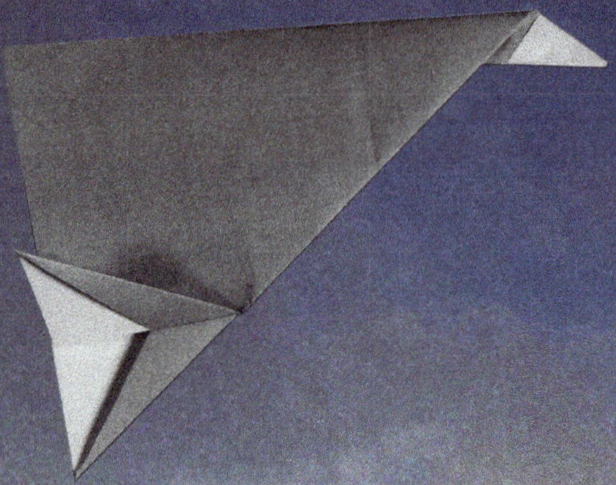

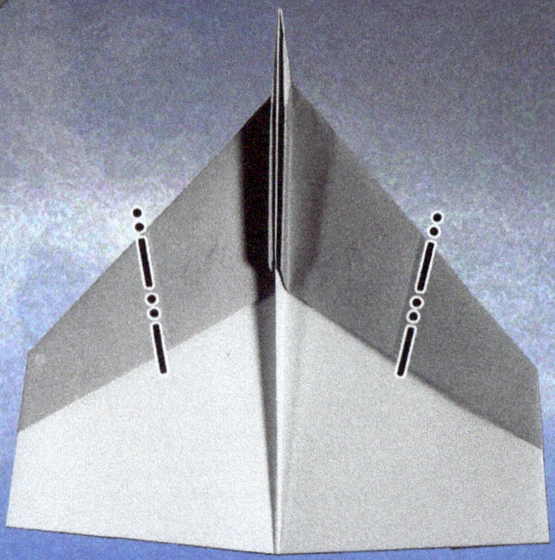

25

Follow the existing creases. Only fold the layered portion of the wings.

STAR FIGHTER

Featured in the move "Paper Planes", this international distance winner adds muscle and glitz to your lineup.

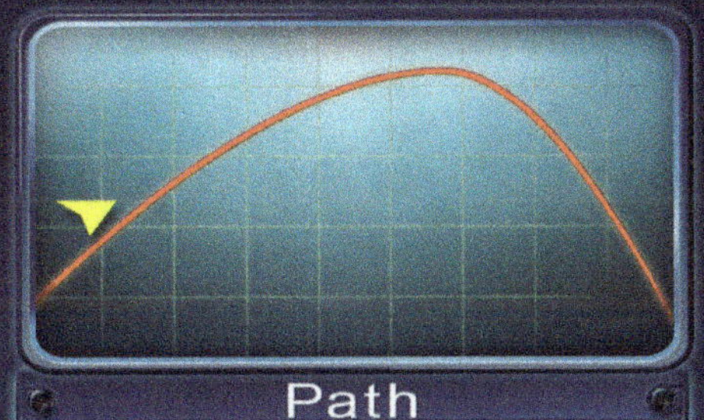

This plane gives you a great reason to learn one of the best base folds in paper airplane making:
the waterbomb base.

Star Fighter Steps

1

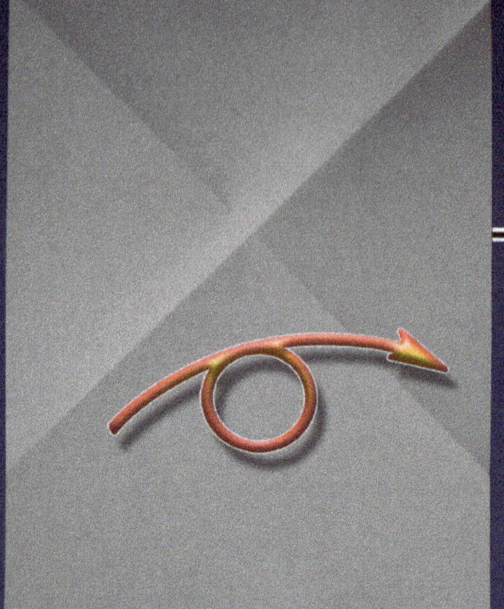

Make diagonal folds and then flip the paper over.

2

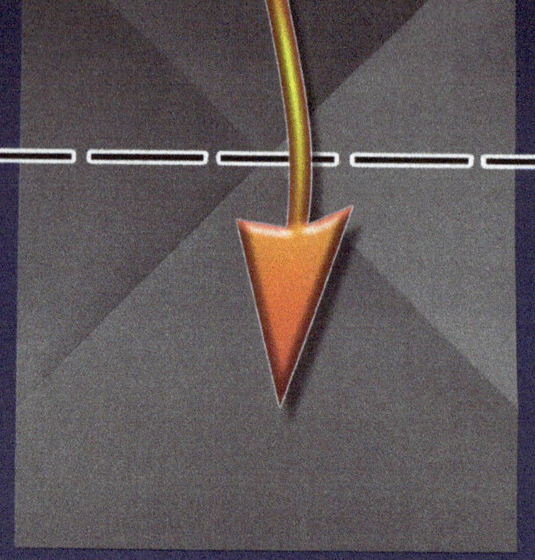

Fold across the center of the X.

3

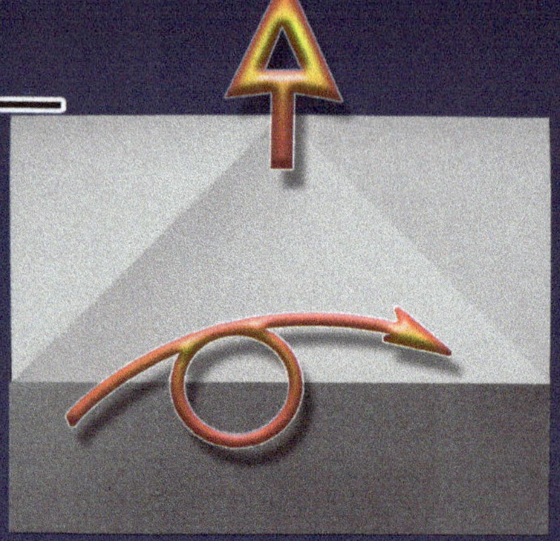

Partially unfold 2 and flip the page over. Make sure the X stays on the top half of the page.

4

This is the fun part. Press at the center until the paper "pops".

5

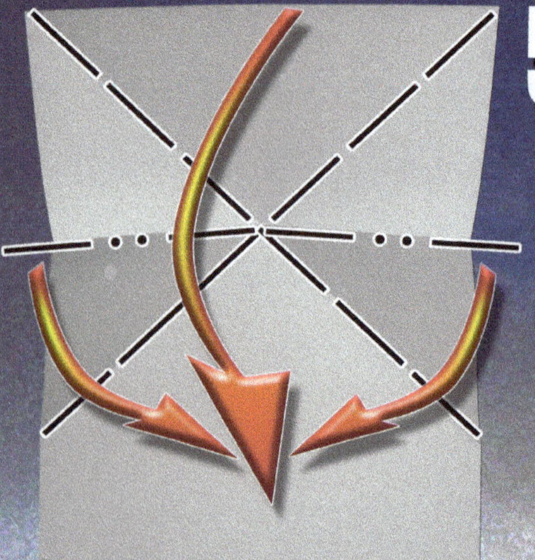

The sides come in as the top comes down. If your creases are sharp and accurate, this happens easily.

The waterbomb base. An amazing start from just three creases.

Star Fighter Steps

6

Fold the right top corner to the left. Notice how the result makes a small triangle which matches the shape of the waterbomb top layer.

7

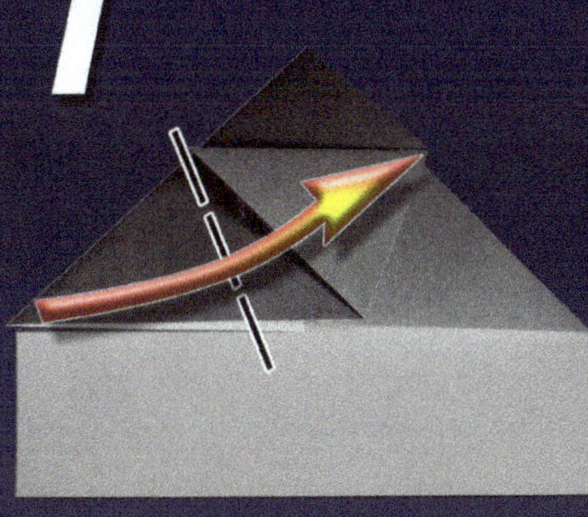

Move the left corner over in the same way. You may need to adjust 6 and 7 so corners line up neatly.

8

Once the layers line up nicely, tuck the right flap inside the left flap.

9

Here's that flap going in.

10

Flip the plane over.

11

Fold the top triangle from the other side down.

11 ¼

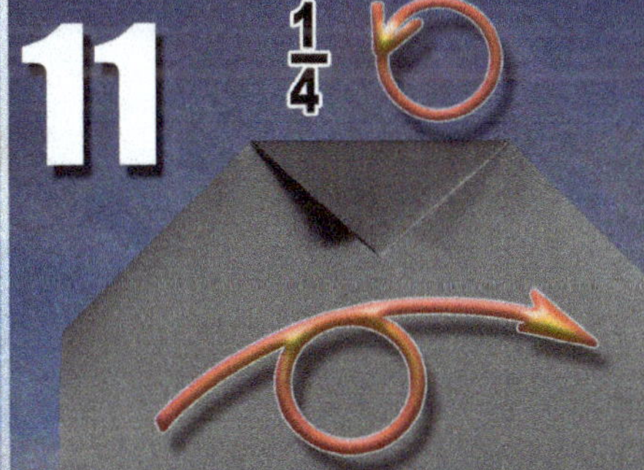

Flip the plane over and rotate the plane left.

12

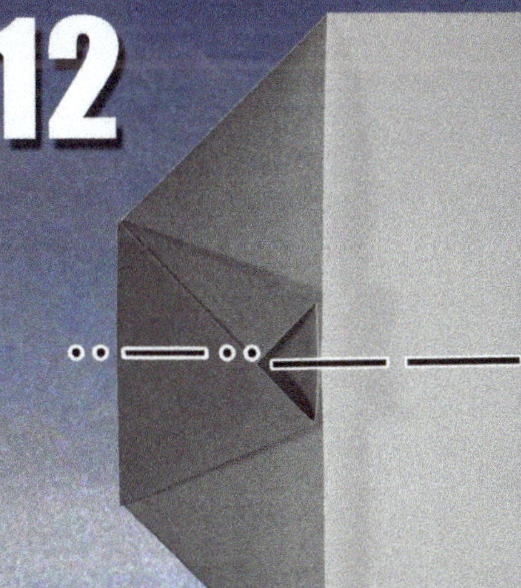

This is where things get weird. Start forming the tunnel by lifting the interlocked flaps, and valley folding the other layers.

The next step shows this in progress.

13

The layers from step 12 half way there.

14

Pinch the top of the tunnel together and push it down. The next step shows the move happening.

15

A fold will form as you keep pushing the tunnel down .

16

As the crease forms, keep it parallel to the top of the tunnel. Also, notice how the crease ends at the rear corner of the interlocked flaps.

17

The wings are now upright, and the tunnel is tucked in neatly. Make a wing crease that lines up with the tunnel crease. Both the wing and tunnel crease should be parallel with the center crease.

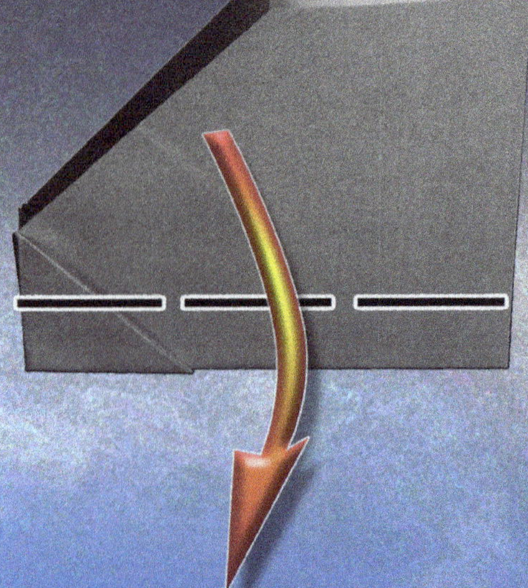

Star Fighter Steps

18

Make a crease parallel with creased edge, using the corner of the tunnel as the starting point.

19

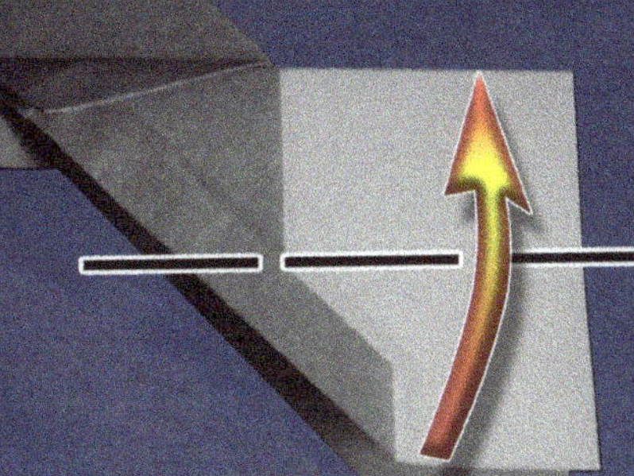

Unfold step 18 enough to reverse fold the point in front.

20

See how the reversed point lets that corner stick out? Good. Now repeat 17-19 for the other side.

21

Fold the bottom edge up to meet the top.

22

Unfold step 21

23

Fold the bottom to the crease.

24

Repeat steps 21-23 for the other wing.

Adjust the wings to look like this.

TWIN JET

The look is based in part on the F-14 and F-18 jet engine placement.

This will test your detail skills. A sink fold gets added to your tool kit. Do you feel the need for speed?

Twin Jet Steps

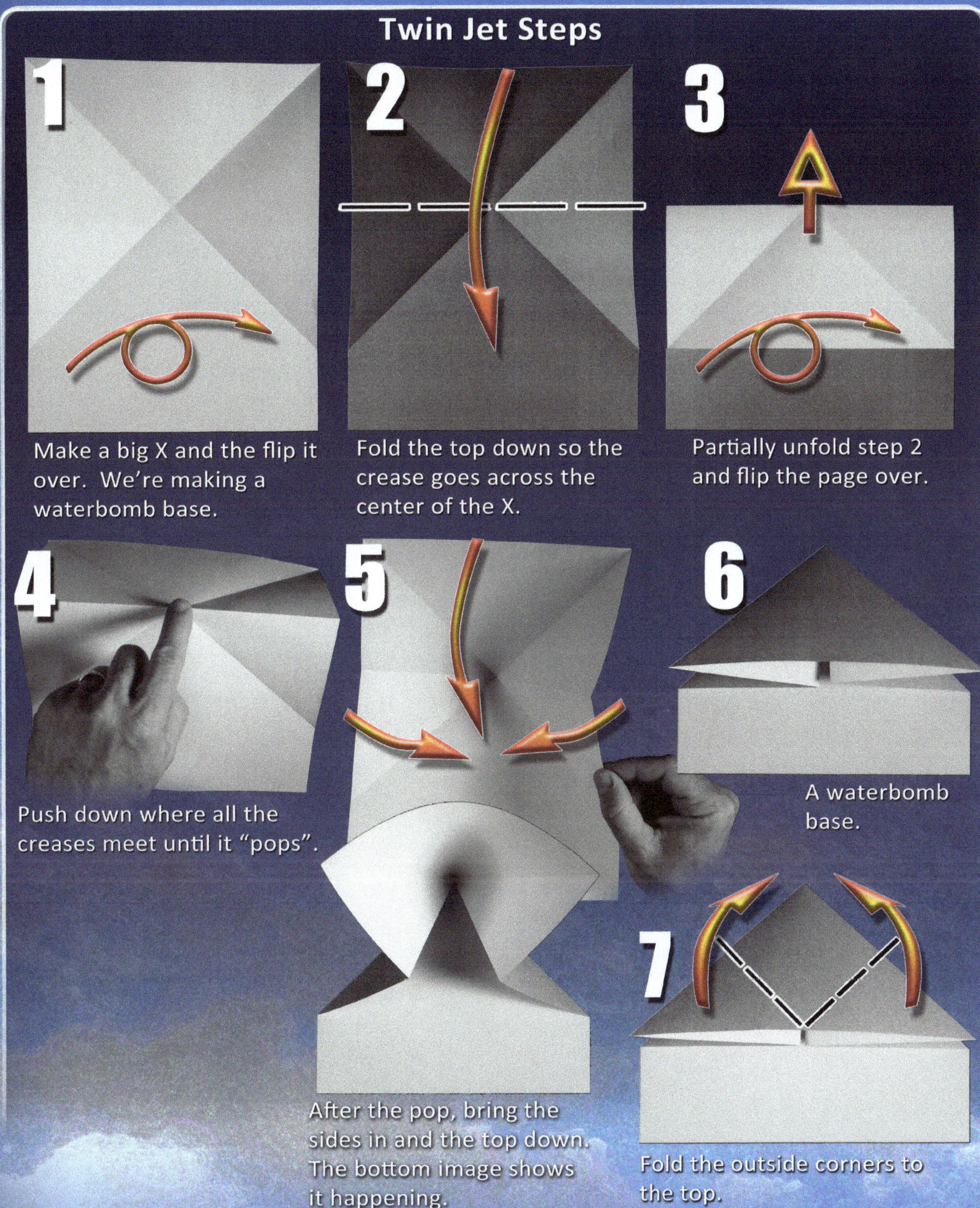

1 Make a big X and the flip it over. We're making a waterbomb base.

2 Fold the top down so the crease goes across the center of the X.

3 Partially unfold step 2 and flip the page over.

4 Push down where all the creases meet until it "pops".

5 After the pop, bring the sides in and the top down. The bottom image shows it happening.

6 A waterbomb base.

7 Fold the outside corners to the top.

8

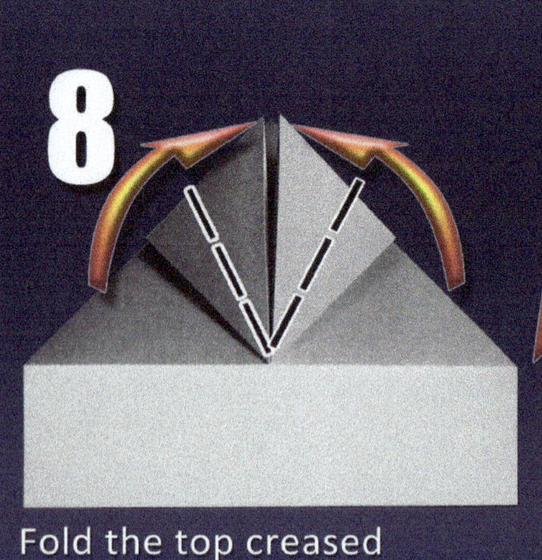

Fold the top creased edges to the center.

9

Unfold 7-8

10

Fold the top down to the lower point. The crease hits the top layer corners.

11

Unfold step 10

12

Fold the top down to the center of the crease from step 10

13

Unfold step 12

14

Get ready for a sink fold by unfolding everything. Flip the paper over.

15

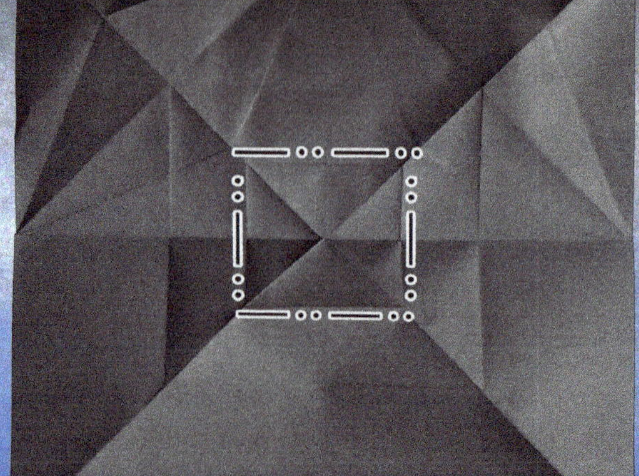

The lower image is zoomed in. Go around the small square with mountain folds.

16

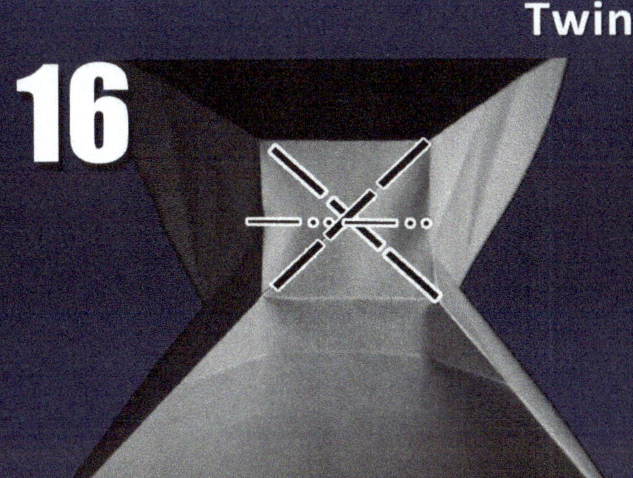

We're turning that little square into a waterbomb base that points down.

17

The sink fold is complete. Flip the plane over.

Notice the combination of mountain and valley folds. Press the center down and make the sides come together. This shows the move about half way there.

18

Follow the existing crease to pull the layers down. The small waterbomb base gets flattened out.

19

Flip the plane over.

20

Fold the top creased edges in, making the crease parallel to the edges. The top edges go to the center crease, making a point.

21

Follow an existing crease to move this flap to the left.

22

Use another existing crease to move the flap right.

23

Repeat 21-22 for the left side.

24

Flip the plane over.

25

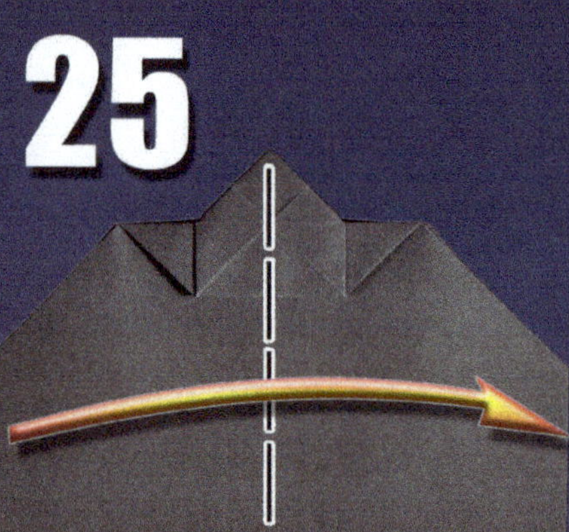

Fold the plane in half.

26

Pull the flaps over.

27

Make a wing crease that is parallel with the center crease. It should just cut through the top layer triangle's right corner.

28

Make the other wing match.

29

Unfold 27-28

30

Fold the flap over where
the crease meets the top.

31

Repeat 29 for the other
side.

32

Wrap the top flap over
and tuck it between
the layers.

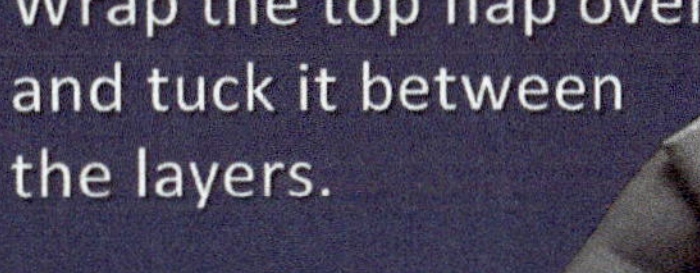

33

Flatten the engine.
A new crease forms
against the wing as you
follow the crease at the
bottom of the fuselage.

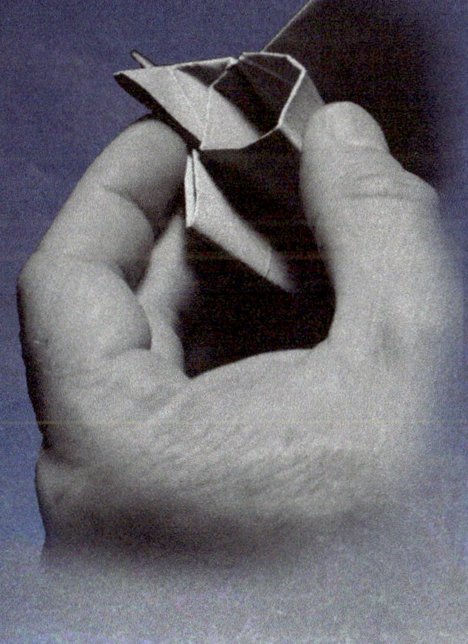

34

Remake the wing crease to finish
squaring up the jet engine. Take
care to keep the point tucked in.
It can easily work loose.

And repeat 31-32 for the other
engine.

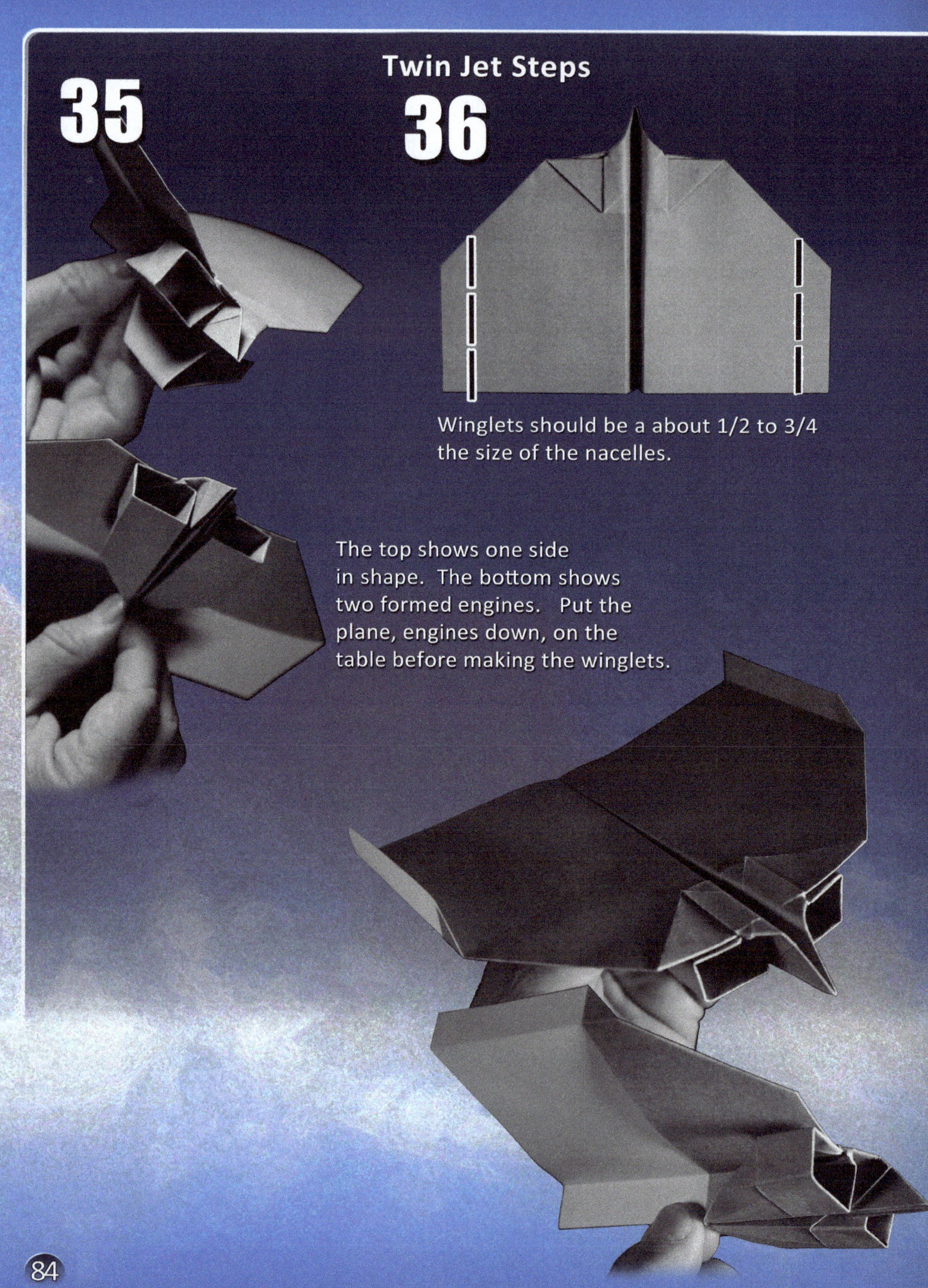

Twin Jet Steps
35
36
Winglets should be a about 1/2 to 3/4
the size of the nacelles.

The top shows one side
in shape. The bottom shows
two formed engines.   Put the
plane, engines down, on the
table before making the winglets.

BONUS PLANE

Glider

I have great memories of this plane catching thermals and flying beyond my visual range circa 1969-1972.

The average indoor flights are misleading. The real fun here is outdoors. It takes luck, the right conditions, and a whole bunch of throws to get a flight to go out of sight.

Bonus Plane Steps

1

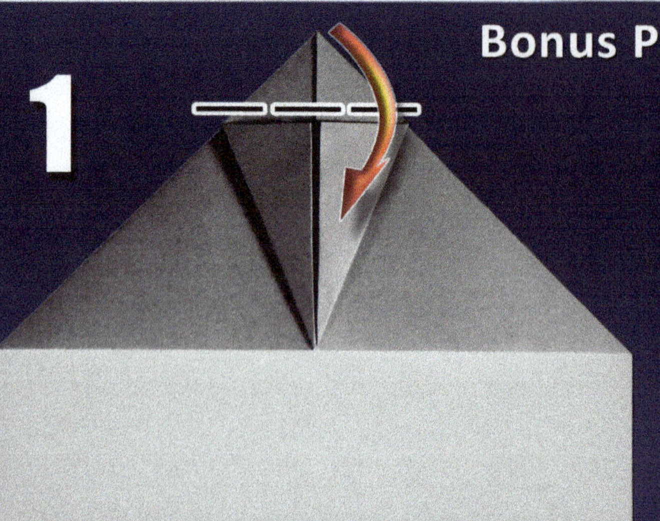

Fold like the Twin Jet through step 8. Then fold the top triangle down.

2

Lift the triangle just enough to put a corner into the slot.

3

Repeat step 2 for the left side.

4

Flip the plane over.

5

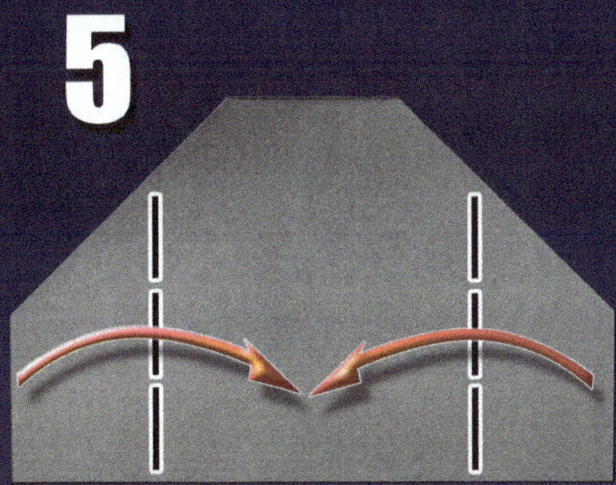

Fold the outside edges to the center. Use the center of the layers from the other side to help line this up.

6

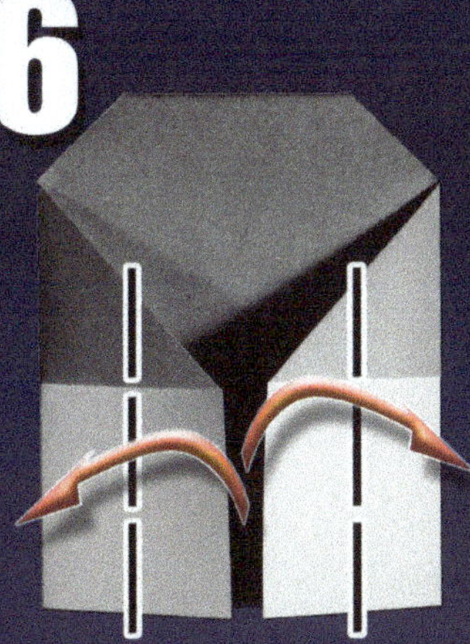

Fold the edges back to the creases from step 5.

7

Crank in a lot of up elevator. Adjust wings so all the angles are square.

Try throwing from the quiet side of a building on a windy day. Launch straight up and try to get into the air coming over the building. It's a good way to catch a long ride.

INTERLOCK DART

Solid, locked in nose weight and enough wingspan for added distance means you'll wow the crowd every time.

A modified waterbomb base and a weird sink fold move will challenge your folding skills. It's worth it. It's one of the meanest, fastest darts you can fold.

Interlock Dart Steps

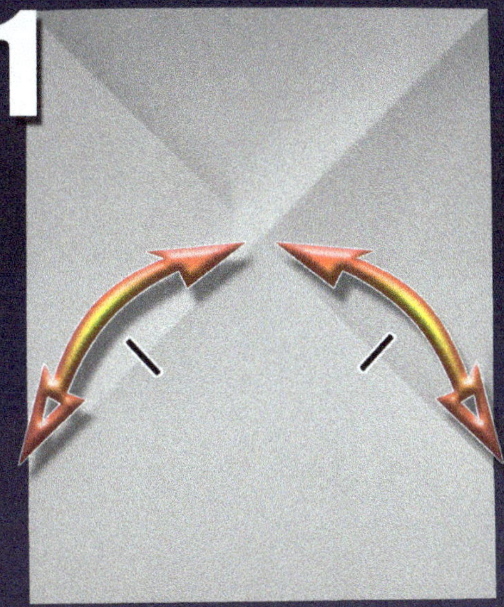

1

Start by making diagonal folds. Then mark the mid-points of the lower legs.

2

To mark a lower leg, move the end of the crease to the center of the X and pinch the crease. Do both sides.

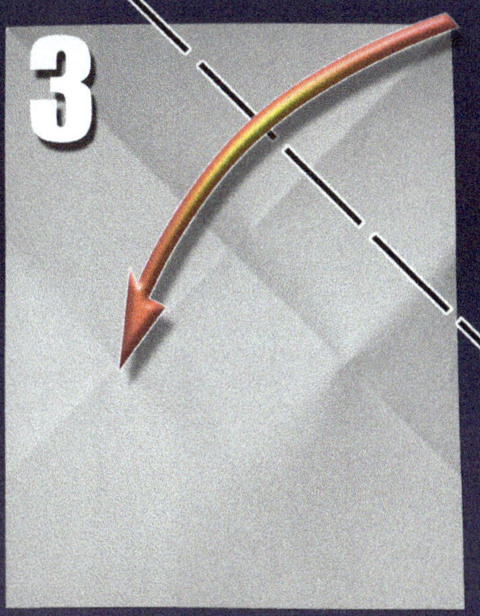

3

Move the upper right corner to the lower left pinch mark.

4

Unfold step 3

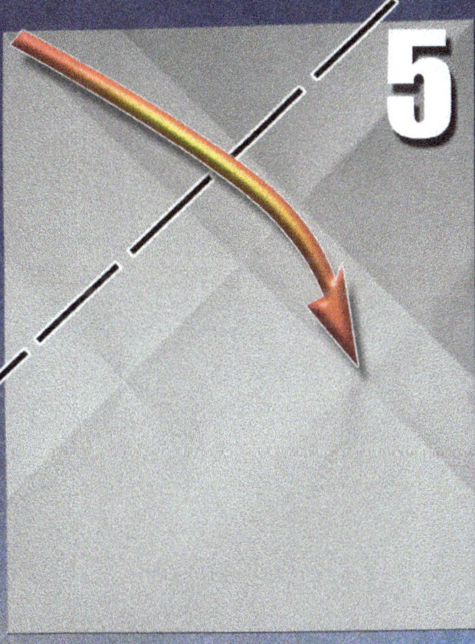

5

Move the upper left corner to the lower right pinch.

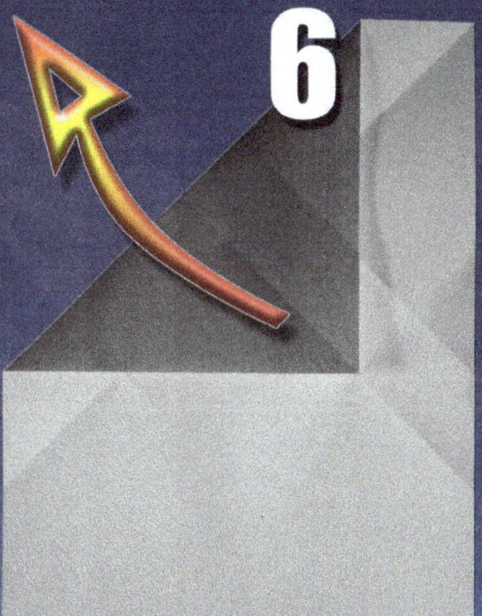

6

Unfold step 5

Interlock Dart Steps

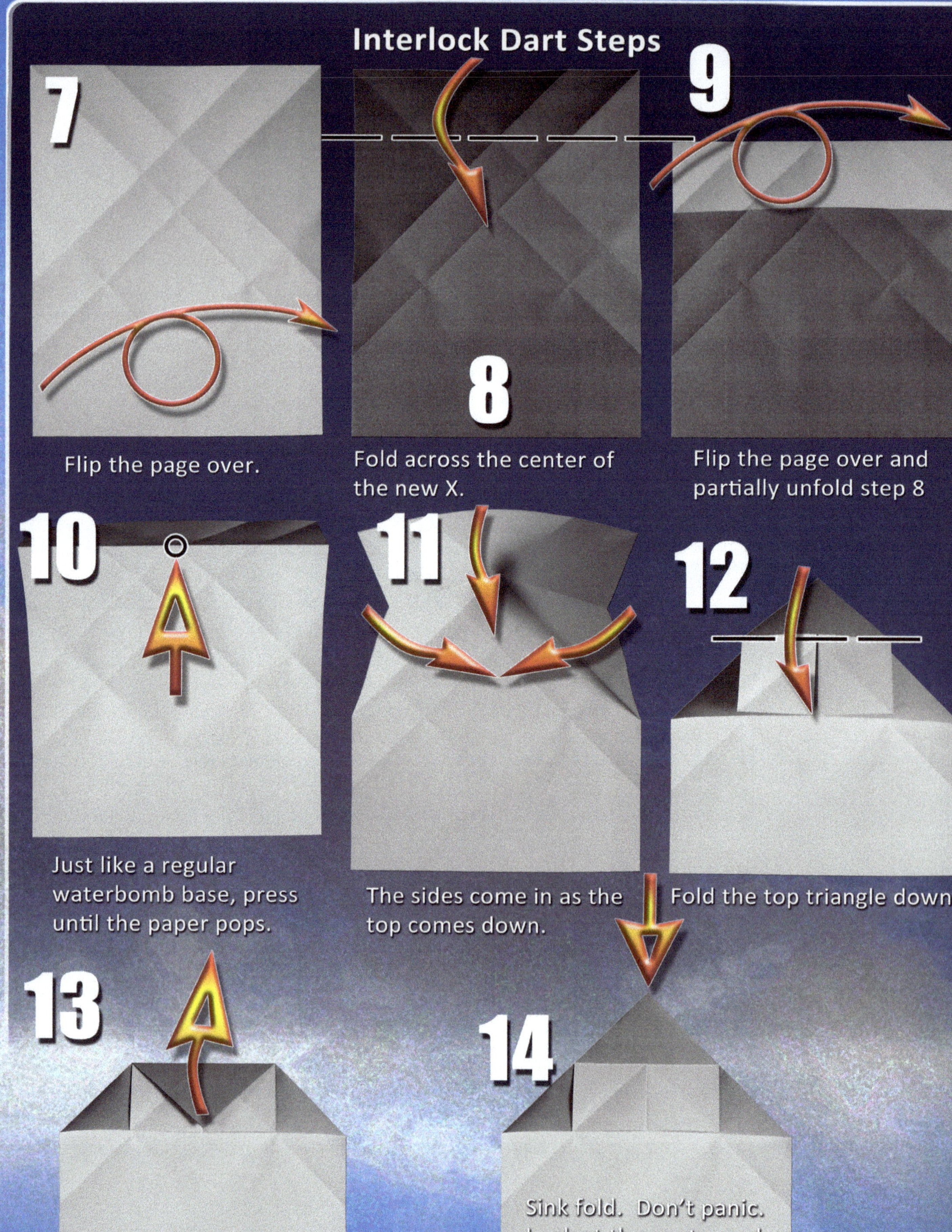

7 Flip the page over.

8 Fold across the center of the new X.

9 Flip the page over and partially unfold step 8

10 Just like a regular waterbomb base, press until the paper pops.

11 The sides come in as the top comes down.

12 Fold the top triangle down.

13 Unfold step 12

14 Sink fold. Don't panic. Look at the next couple steps.

Interlock Dart Steps

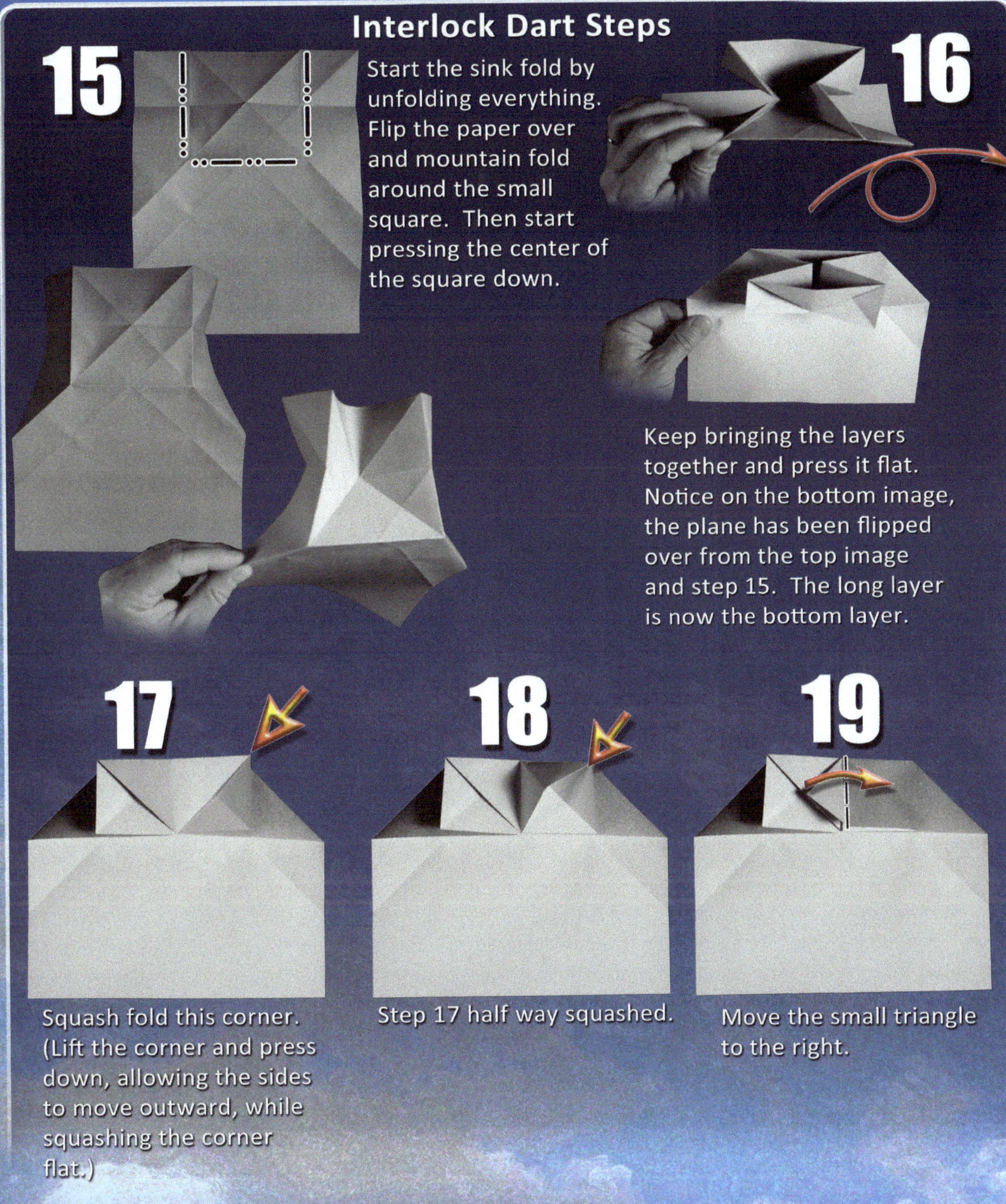

15

Start the sink fold by unfolding everything. Flip the paper over and mountain fold around the small square. Then start pressing the center of the square down.

16

Keep bringing the layers together and press it flat. Notice on the bottom image, the plane has been flipped over from the top image and step 15. The long layer is now the bottom layer.

17

Squash fold this corner. (Lift the corner and press down, allowing the sides to move outward, while squashing the corner flat.)

18

Step 17 half way squashed.

19

Move the small triangle to the right.

Interlock Dart Steps

20

Repeat 17-19 for the left side.

21

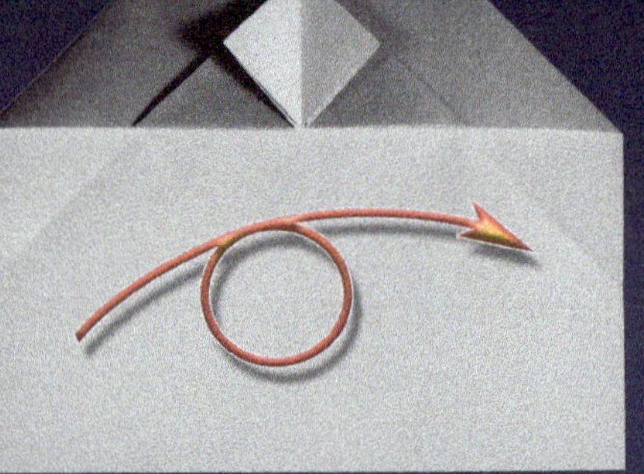

Flip the plane over.

22

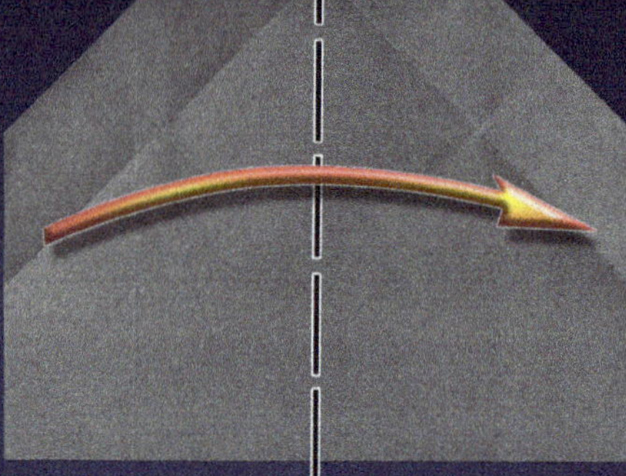

Fold in half.

23

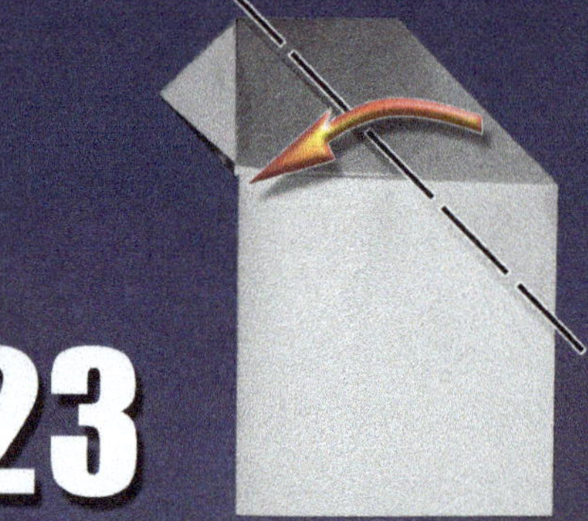

Follow the old crease.

24

Open up the pocket and put the triangle in there.

25

Step 24 looks like this when it's happening.

26

Repeat 24 for the other side.

27

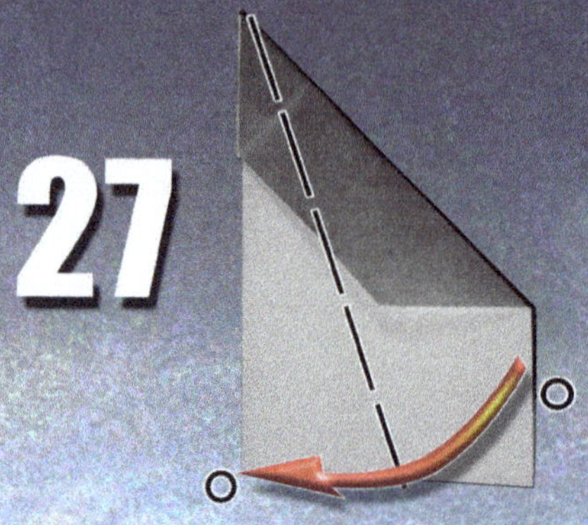

The wing fold starts at the nose. Make the crease when the raw edge just touches the marked corner.

28

Repeat 27 for the other wing.

INTERLOCK BIPLANE

The unusual two piece construction creates a stunning effect. Outdoor flights are amazing.

Sure, it's complicated: sink, squash, and modified water-bomb base. If you made the Interlock Dart, you can make the Interlock Biplane.

Interlock Biplane Steps

1 Fold one Interlock Dart to step 27. Don't make a wing fold yet.

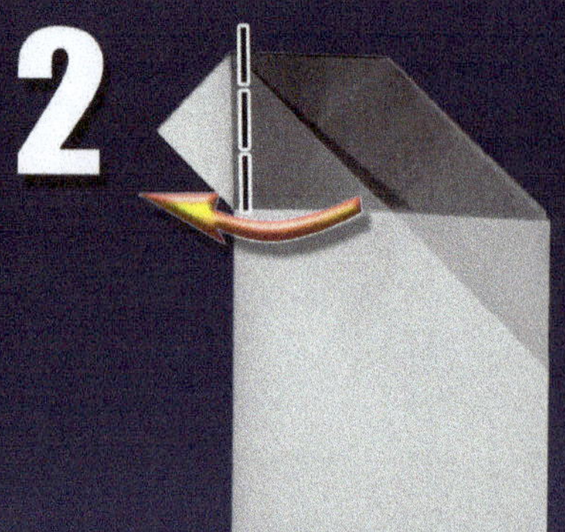

2 Fold a second Interlock Dart. This time up to step 22. Move the bigger triangle to the left.

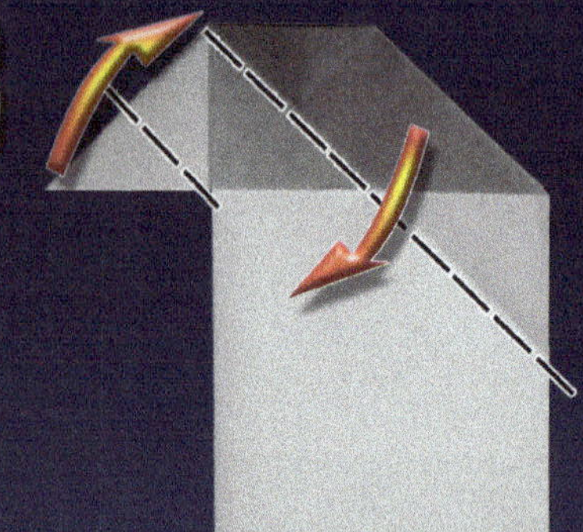

3 Fold the triangle in half. Then follow the existing crease.

4 Put the top triangle only into the pocket.

5 Once this side is in cleanly, do the same thing to the other side (2-4).

6 Make wing folds that fold the pocket in half.

7 Unfold the wings and rotate the plane to the right. The tabs go into the pockets on step 1.

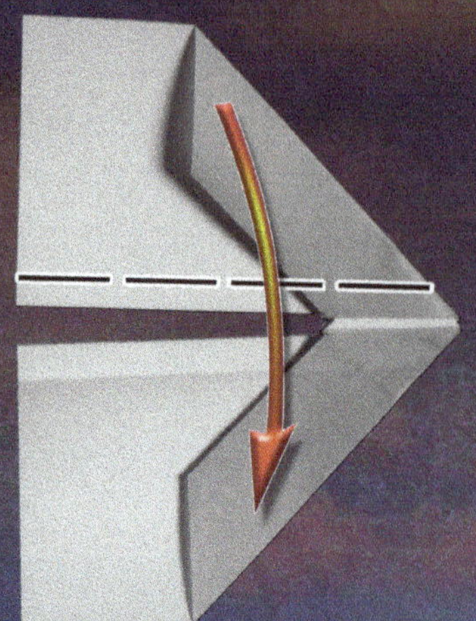

8 Make the wing fold on the upper half. The same place as step 6.

9 Repeat 8 for the other topside wing.

10

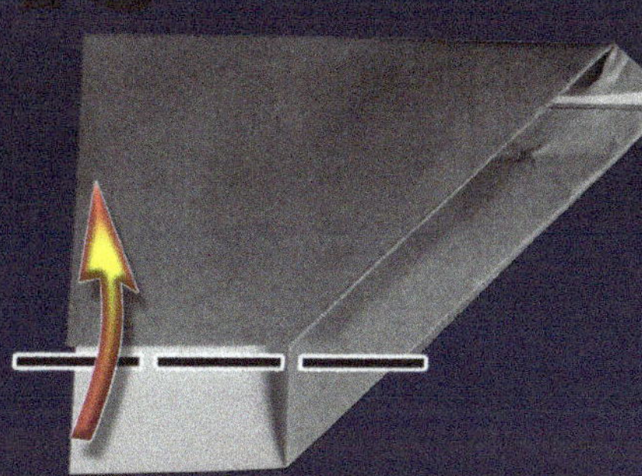

Fold the lower wing up using the upper wing as a guide.

11

Repeat 10 for the other side.

12

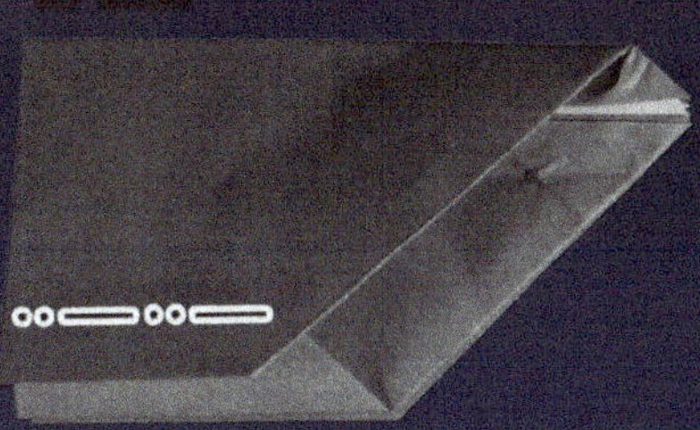

Make downward pointing winglets on the upper wing, about 1/2 the height of the lower winglets

You can add up elevator to the upper or lower wings. Outside, the Interlock Biplane can ride the wind in wild ways.

Sometimes it crabs sideways, and other times it just takes off on long, smooth flights.

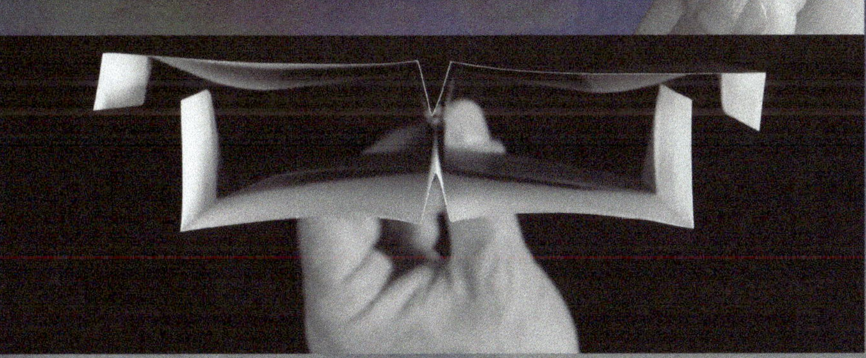

BOOMERANG I

I'm always asked, "What's your favorite plane?" My answer is "for what?" This is my favorite stunt plane.

It's challenging to fold, and the flying is phenomenal. The Boomerang I circles either direction and loops!

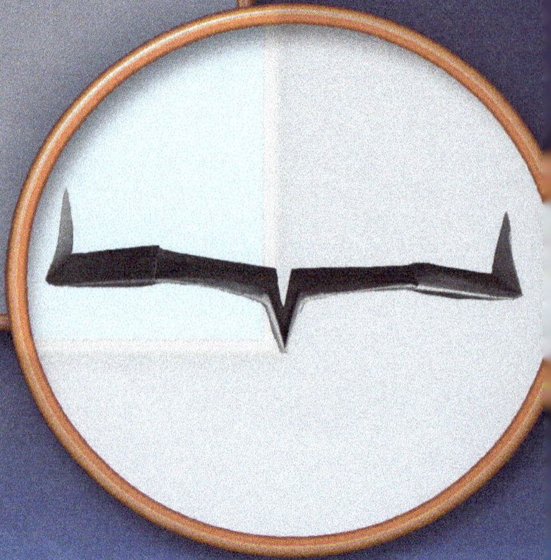

Boomerang I Steps

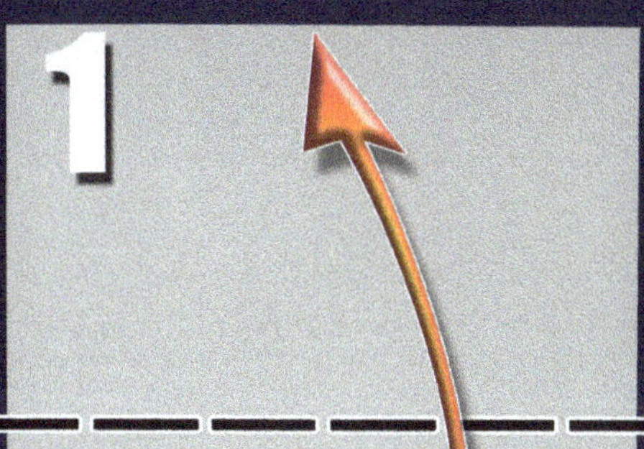

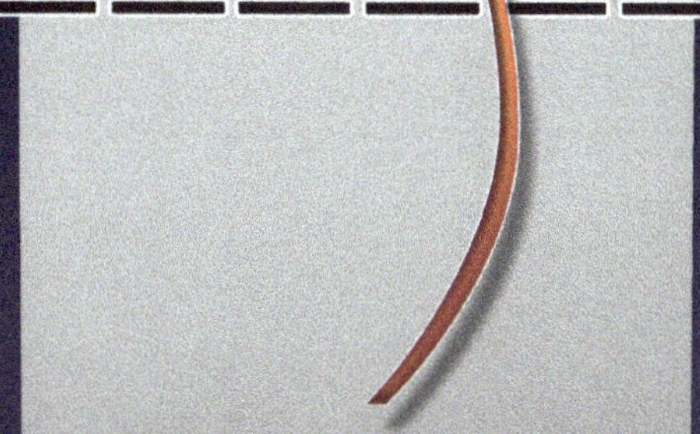

1 Start with a short side up. Fold the page in half.

2 Mark the half way point of the crease from step 1 with a pinch. Bend one end of the creased edge to meet the other end and press (or pinch) the center flat.

3 Make a pinch at the half way point between the step 2 pinch and the left side.

4 Make a fold from the pinch to the upper left corner.

5 Squash this corner.

6 Fold the forward flap behind

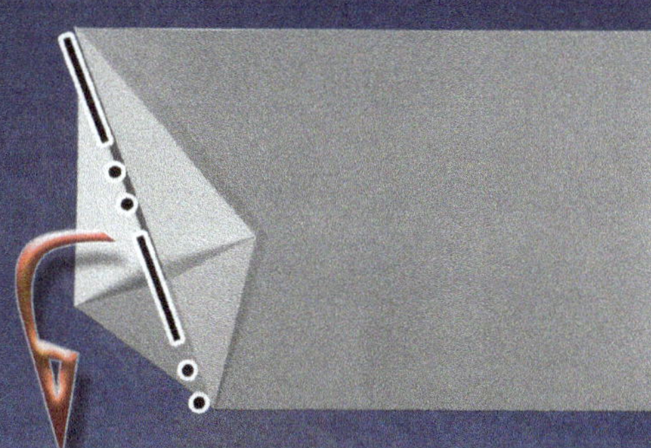

7 Fold the upper corner just short of the lower corner.

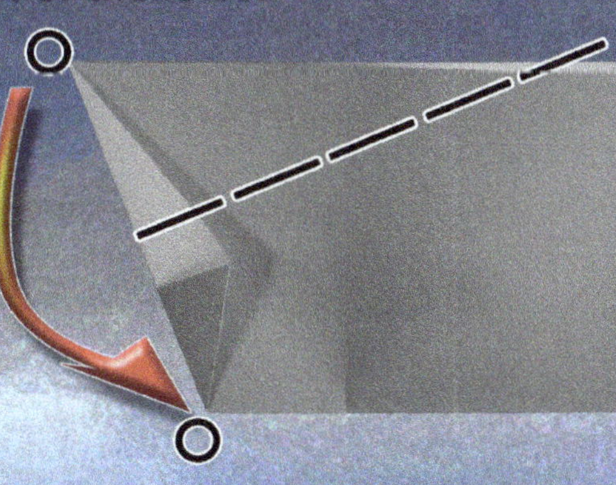

This step is preparation for putting the corner into the pocket fromed by the squash fold from step 5. In addition to landing short of the lower corner, leave a half millimeter of the pocket showing on the front (left side).

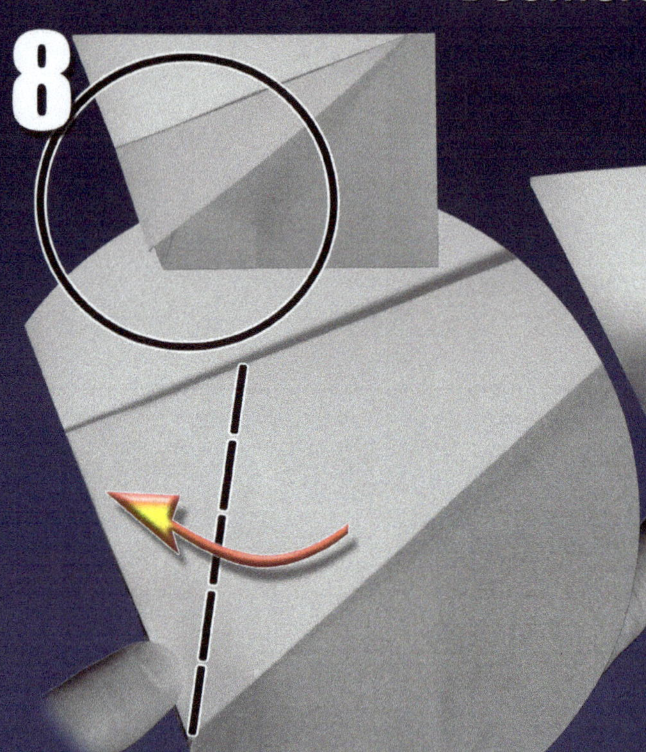

The bottom image is zoomed in. The crease follows the shape of the pocket underneath. Make the crease just to the inside of the pocket so it will fit. Look at the next step.

Two things to notice: The point formed by step 8 will fit inside the pocket formed by steps 5-6, and we're allowing the paper to stay curved until after the point goes into the pocket.

Put the point into the pocket.

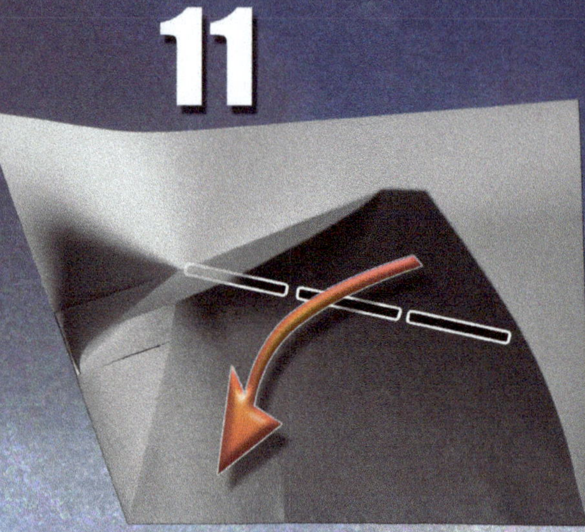

Once the point is cleanly in the pocket, flatten the curved part. A crease starting at the front of the pocket and sloping upward forms as well as the longer marked crease.

12

Note where the raw corner ended up: just touching the center crease.
Repeat steps 7-11 for the other side.

13

Use the corners as guides for this crease, on both sides.

14

Push on the pocket to start squashing the whole front. Let the sides move outward.

15

Here's step 14 in progress.

16

Step 14 complete. Now flip the plane over.

17

Fold the corner over to line up with the center crease. The crease goes right where the layers end.

18

Fold the plane in half.

19

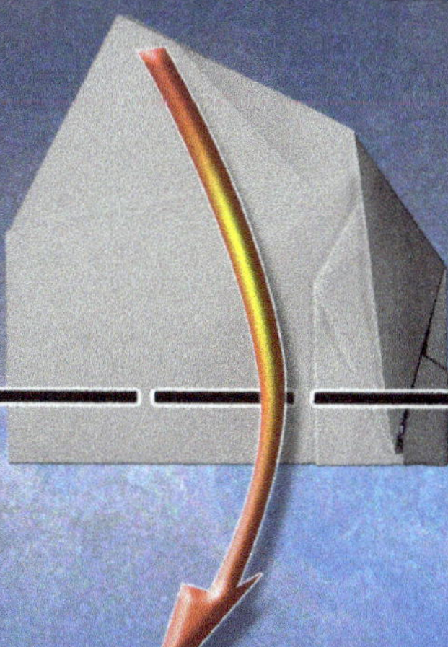

Make the wing crease so it just cuts the tip of the layer up from the center crease on step 18.
Look at the next step to see the result.

20

Notice just a sliver of the triangle gets folded over.

21

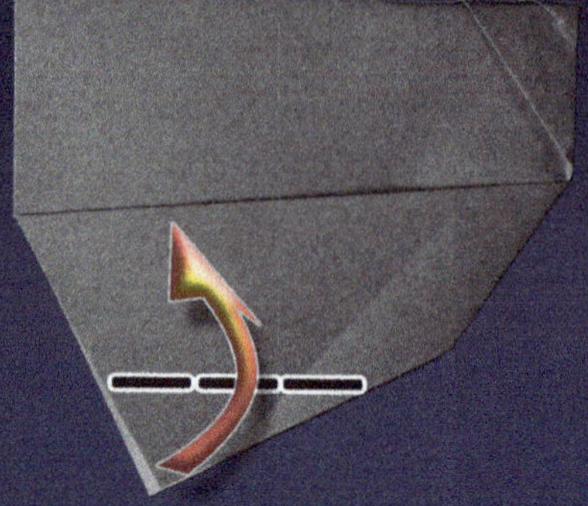

The winglet should be parallel with the wing crease. The raw corner should not go past the layered part.

22

Make the other winglet match.

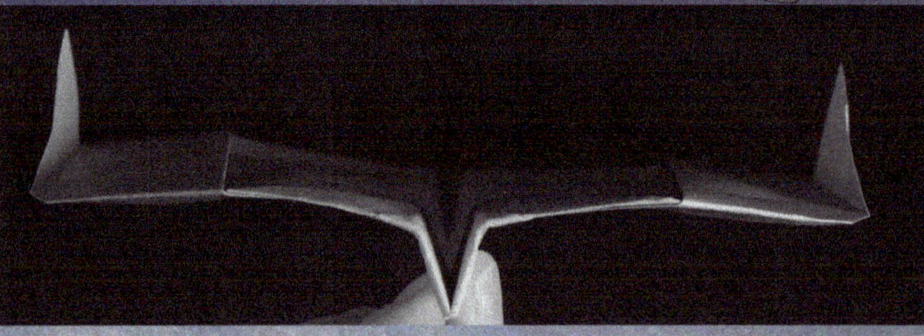

Curve the front of the wings like the picture. Get the plane flying in a straight line before trying to circle. If you can get the plane to loop, it will circle reliably. Circling back is just a loop leaned over. Throw the plane leaned left or right. It will circle either way.

The Boomerang I has drooping wings (negative dihedral or anhedral angle) Planes with positive dihedral easily rock back to neutral when they're launched leaned over. This plane stays leaned over. Because the center of gravity is behind the center of lift, the plane climbs. A climb, when leaned over, is a circle back.

BOOMERANG II

Stunt

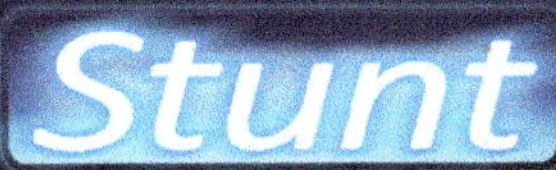

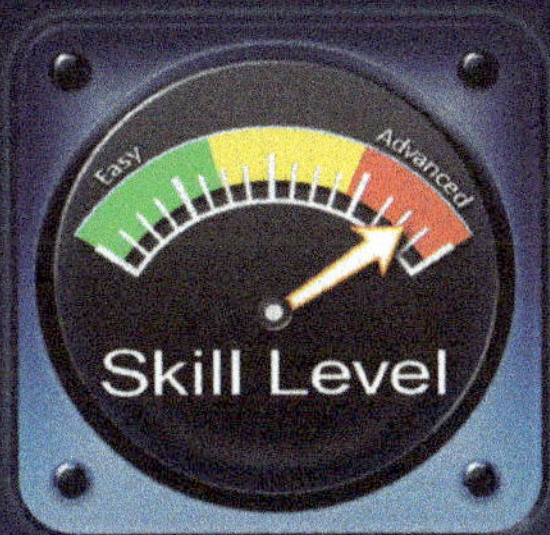

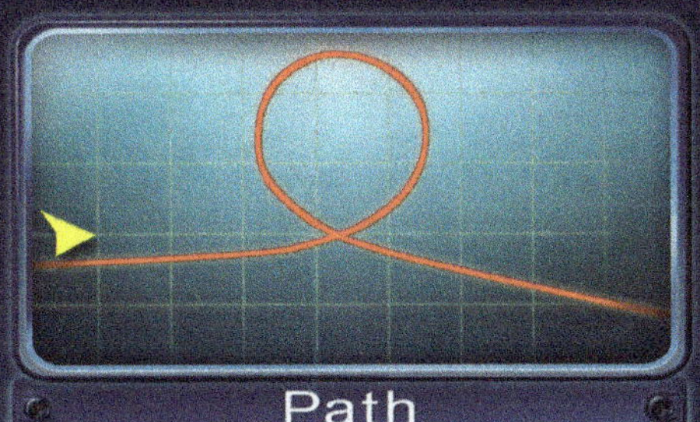

It flies out, flips over, and flies back upside down. That's weird and wonderful.

Yes, it's a folding challenge. If you made it through the Boomerang I, you can do this.

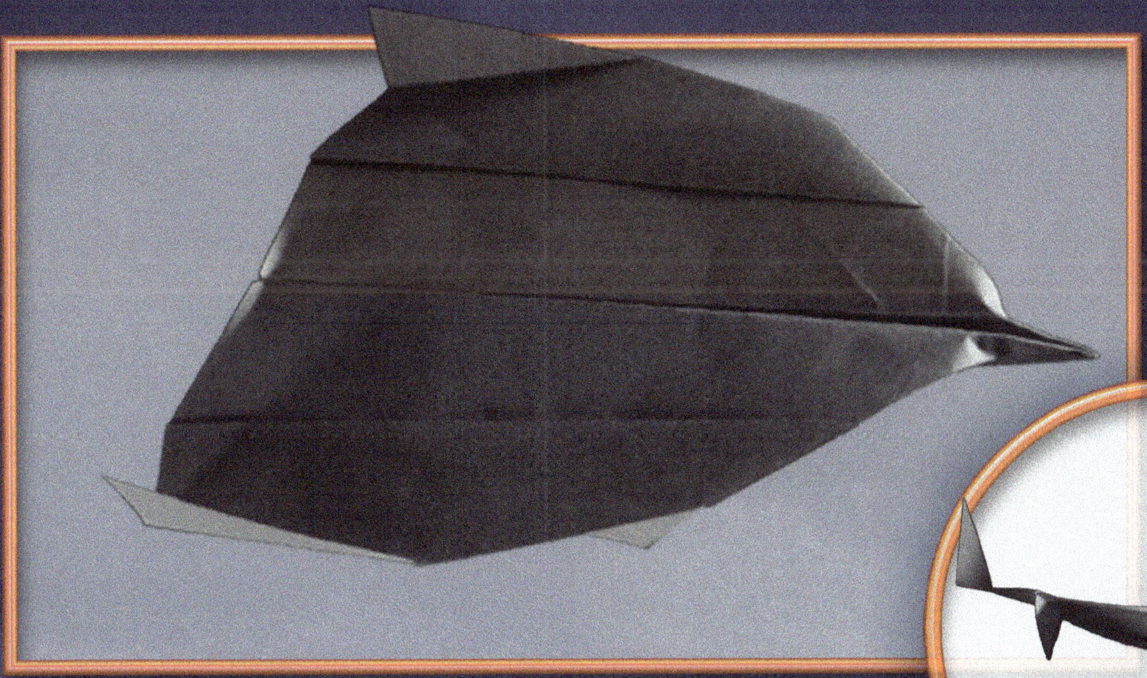

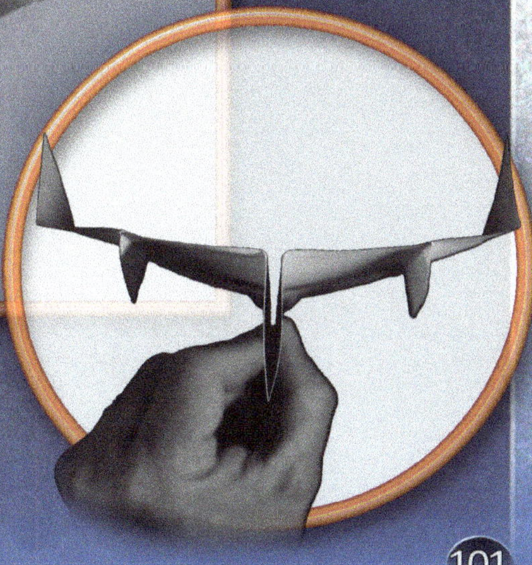

Boomerang II Steps

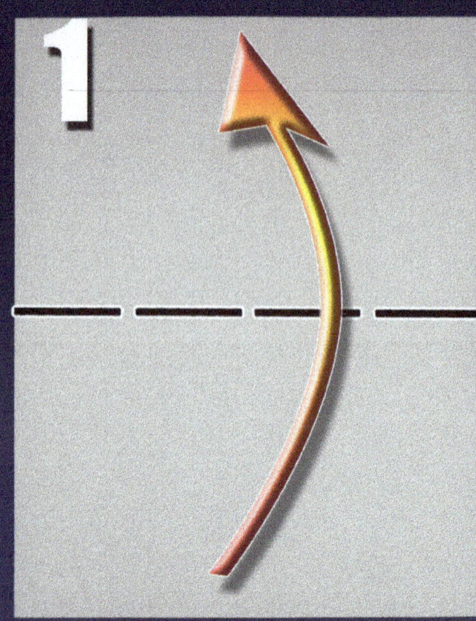

Start with a short side up.
Fold the page in half.

Mark the center of the
step 1 crease with a pinch.

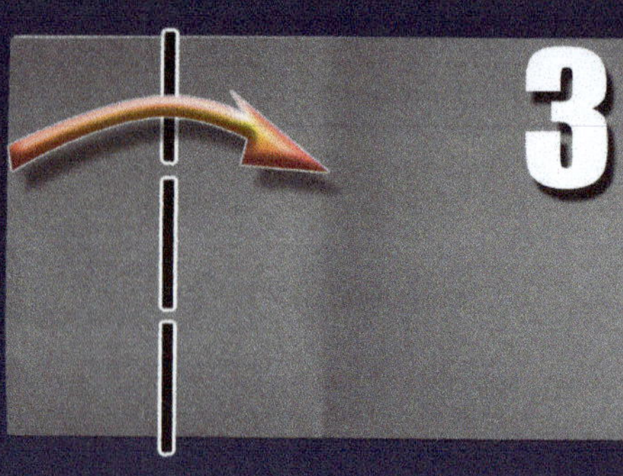

Fold the left edge to the
pinch mark.

Unfold step 3

The crease starts at the
upper left corner. The
marked corner ends up
touching the crease from
step 3

Squash the flap. Open
the layers and push the
point to the center.

Fold the front half of the
squash behind.

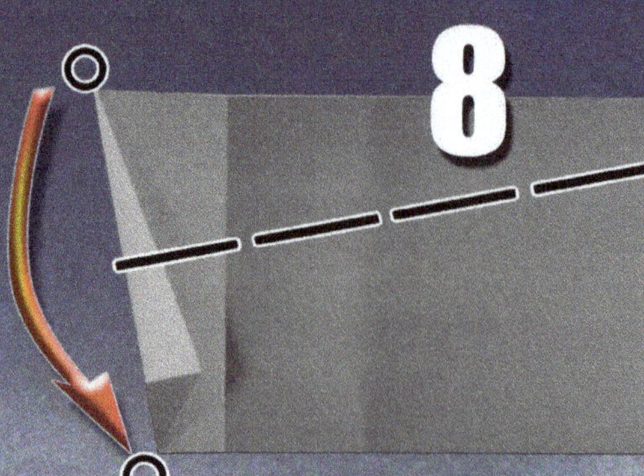

Bring the marked corner
down to just short of the
marked corner of the
pocket.

Start to narrow the
point to fit into the
squash folded pocket.
Fold over so the crease
follows the pocket
below.

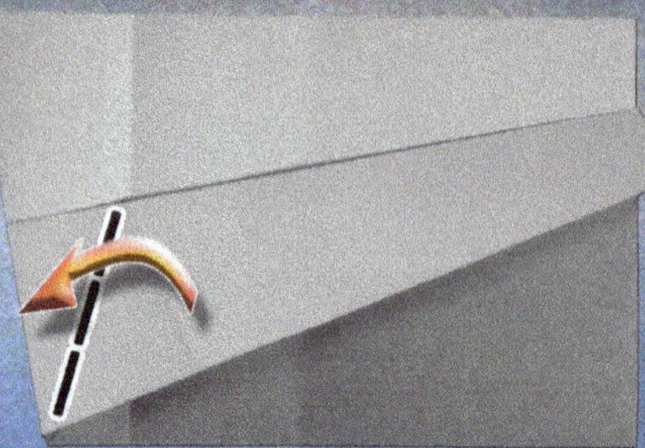

Boomerang II Steps

10

11

Fold up the flap using the corners as natural limits. Of course, you have to do 9-11 to the other side.

12

Press on the front and let the sides move outward. It's like a big squash fold for the front of the plane.

The top image shows the corner on top of the pocket. The middle shows the start of tucking the coner into the pocket. The bottom image is the corner in there. Only now can you flatten the warped layer.

13

Fold the creased edges over to meet the invisible layers. I usually rub my thumbnail along the edge of the hidden layer to create a reference.

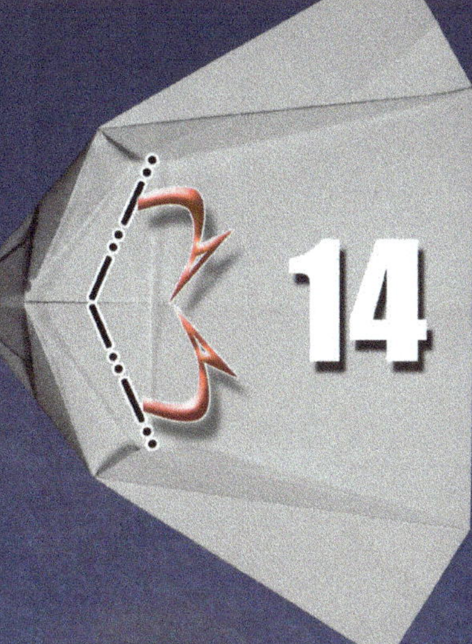

14

Fold these small flaps underneath. It's easier to fold them over to meet the creased edge first, and then reverse the crease to tuck them underneath. I used to say this move was optional, but I haven't skipped it for about 10 years now.

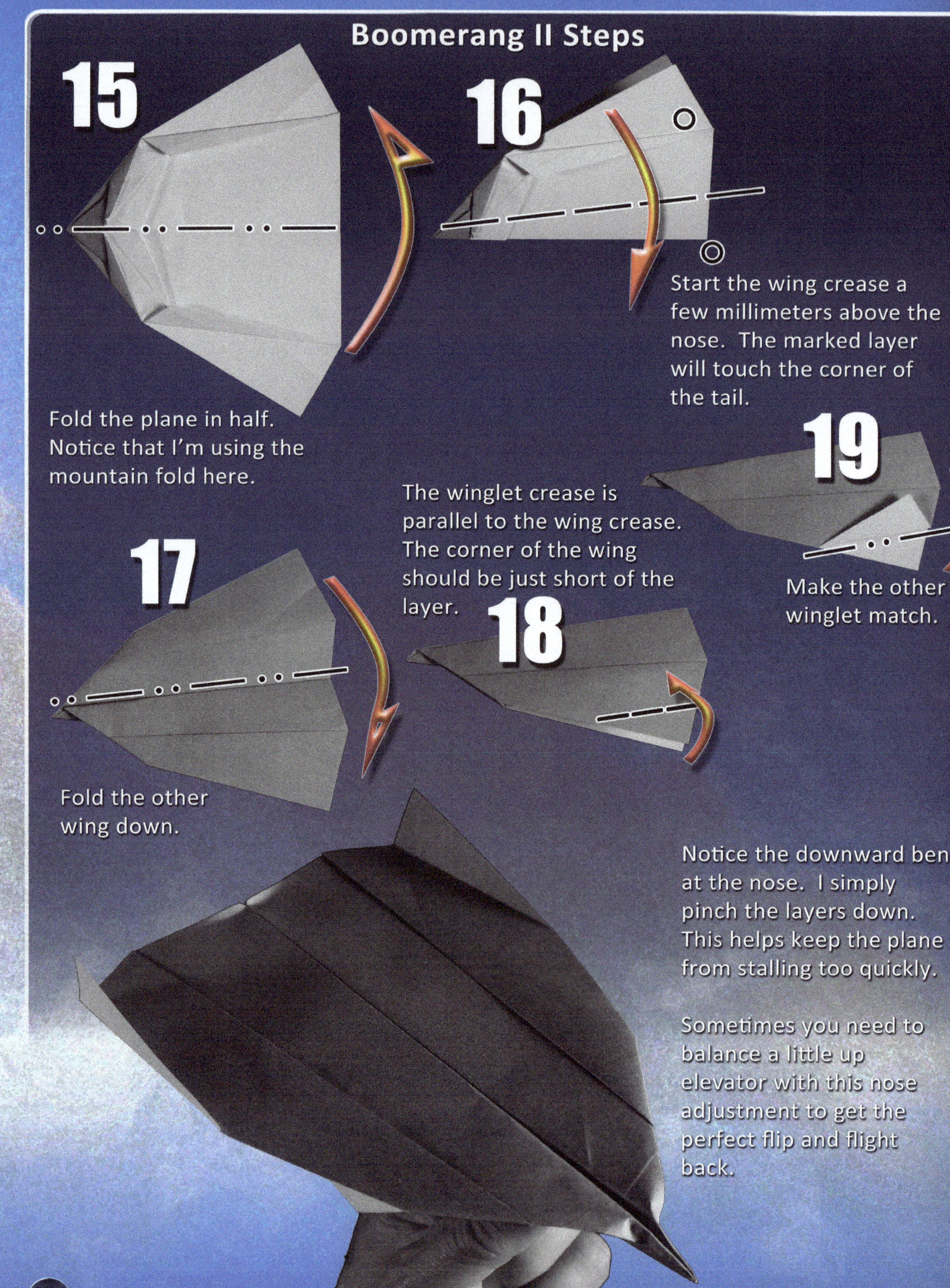

Boomerang II Steps

15

16

Start the wing crease a few millimeters above the nose. The marked layer will touch the corner of the tail.

19

Fold the plane in half. Notice that I'm using the mountain fold here.

The winglet crease is parallel to the wing crease. The corner of the wing should be just short of the layer.

17

18

Make the other winglet match.

Fold the other wing down.

Notice the downward bend at the nose. I simply pinch the layers down. This helps keep the plane from stalling too quickly.

Sometimes you need to balance a little up elevator with this nose adjustment to get the perfect flip and flight back.

FFF-1

Follow Foil

Unlimited flight time is right here. Even loops are possible with a lot of practice. My record is 5 loops in a row.

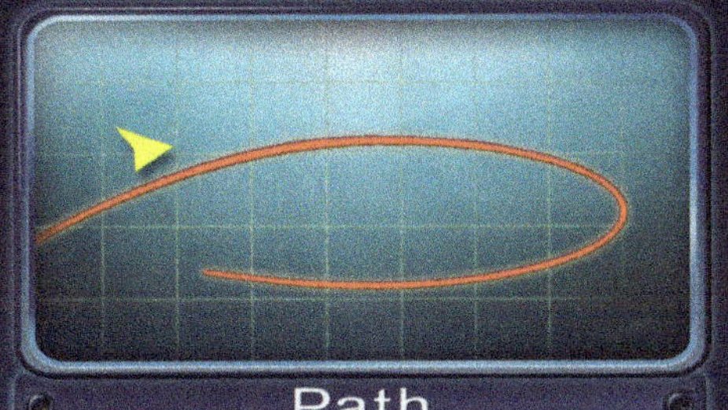

I've flown this kind of plane for more than a half hour. It's so good, Guinness had to change their rules to forbid this kind of plane for duration!

Use 9lb onion skin or phone book paper for the best flights.
(Practice the folding with 20lb paper)

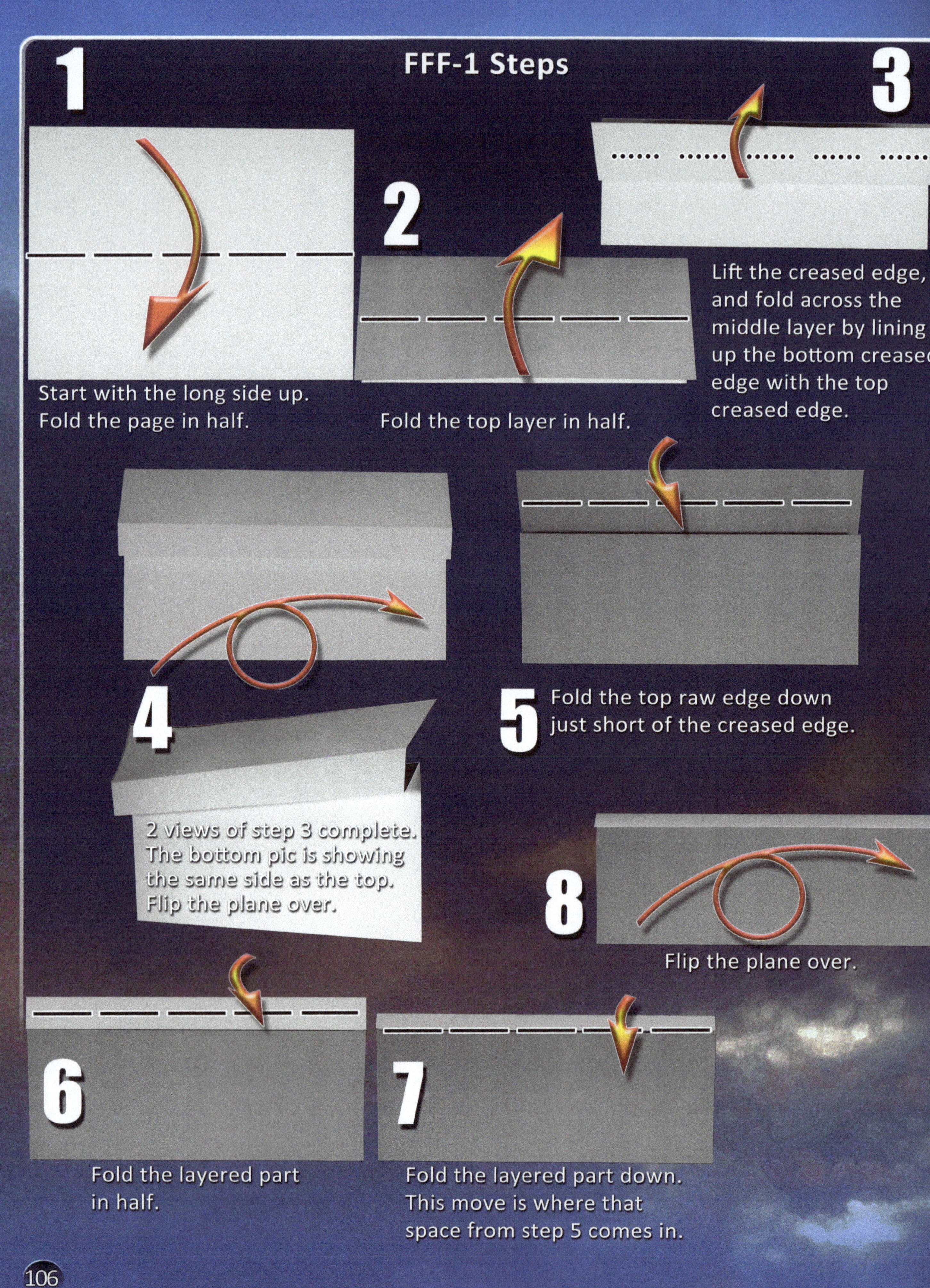

FFF-1 Steps
1
Start with the long side up.
Fold the page in half.
2
Fold the top layer in half.
3
Lift the creased edge, and fold across the middle layer by lining up the bottom creased edge with the top creased edge.
4
2 views of step 3 complete. The bottom pic is showing the same side as the top. Flip the plane over.
5
Fold the top raw edge down just short of the creased edge.
8
Flip the plane over.
6
Fold the layered part in half.
7
Fold the layered part down. This move is where that space from step 5 comes in.

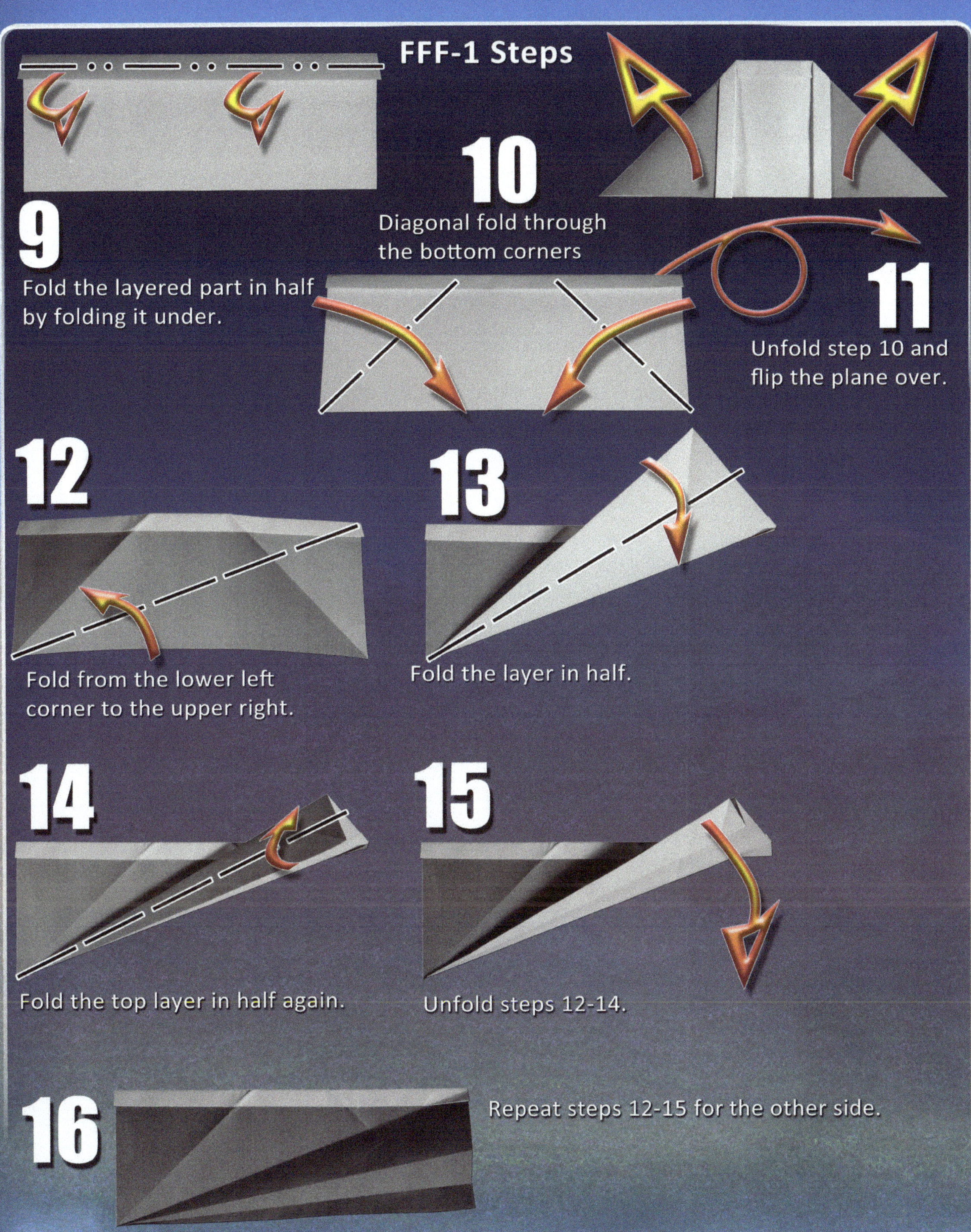

FFF-1 Steps
9
Fold the layered part in half by folding it under.
10
Diagonal fold through the bottom corners
11
Unfold step 10 and flip the plane over.
12
Fold from the lower left corner to the upper right.
13
Fold the layer in half.
14
Fold the top layer in half again.
15
Unfold steps 12-14.
Repeat steps 12-15 for the other side.
16

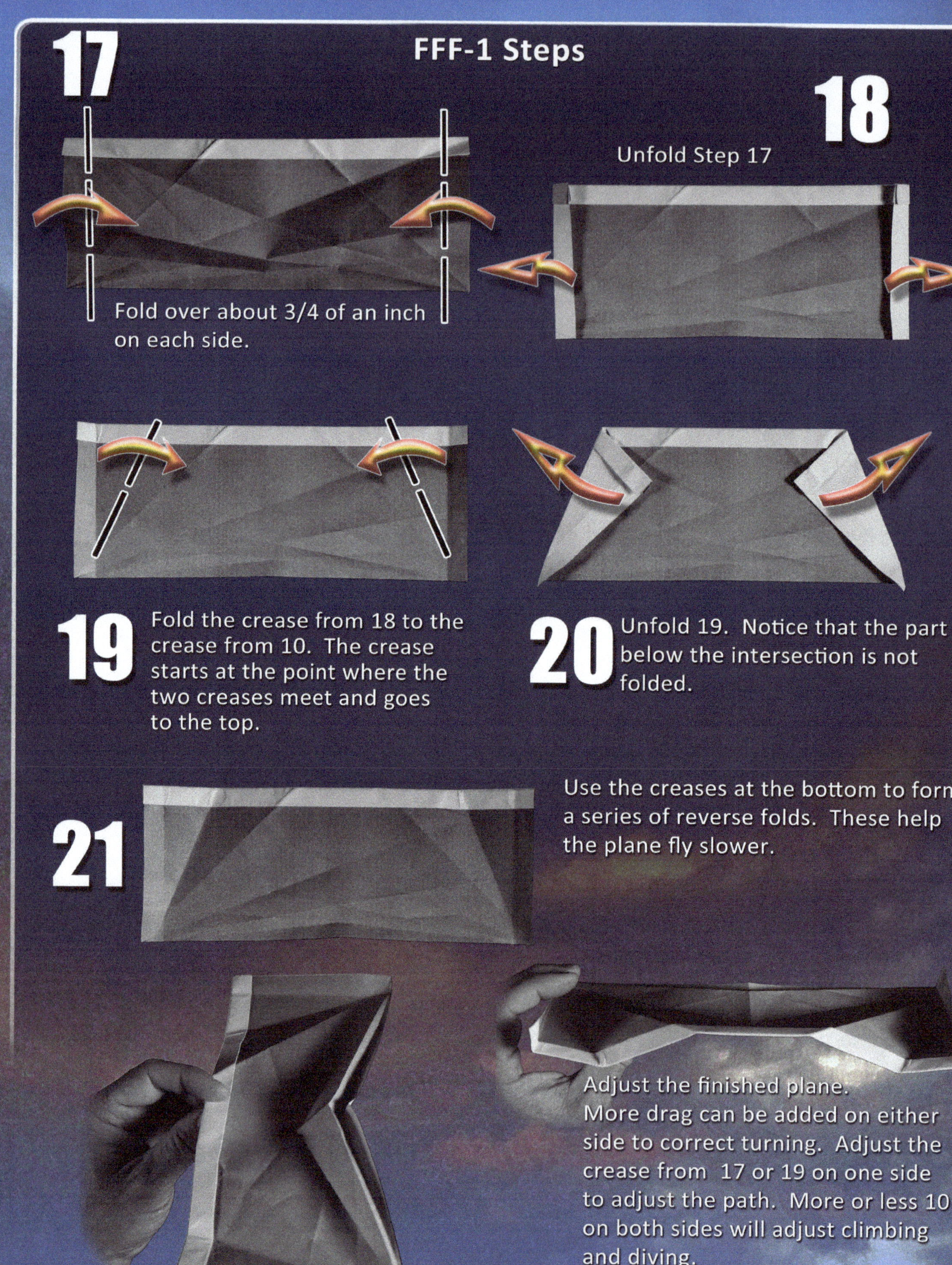

17

Fold over about 3/4 of an inch on each side.

18

Unfold Step 17

19 Fold the crease from 18 to the crease from 10. The crease starts at the point where the two creases meet and goes to the top.

20 Unfold 19. Notice that the part below the intersection is not folded.

21 Use the creases at the bottom to form a series of reverse folds. These help the plane fly slower.

Adjust the finished plane. More drag can be added on either side to correct turning. Adjust the crease from 17 or 19 on one side to adjust the path. More or less 10 on both sides will adjust climbing and diving.

THE PELICAN

Glider

This plane was first made and flown in 1986, when I was writing The Gliding Flight.   Origami and performance meet.

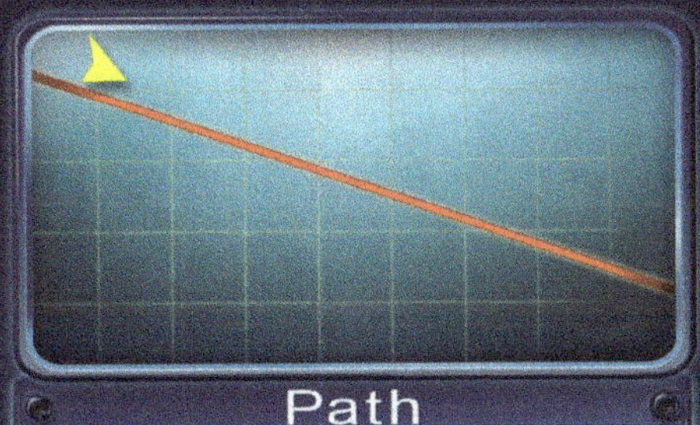

It's a very good glider. Try adding even more up elevator to fly some loops. The Pelican can handle high wind and stunts.

Pelican Steps

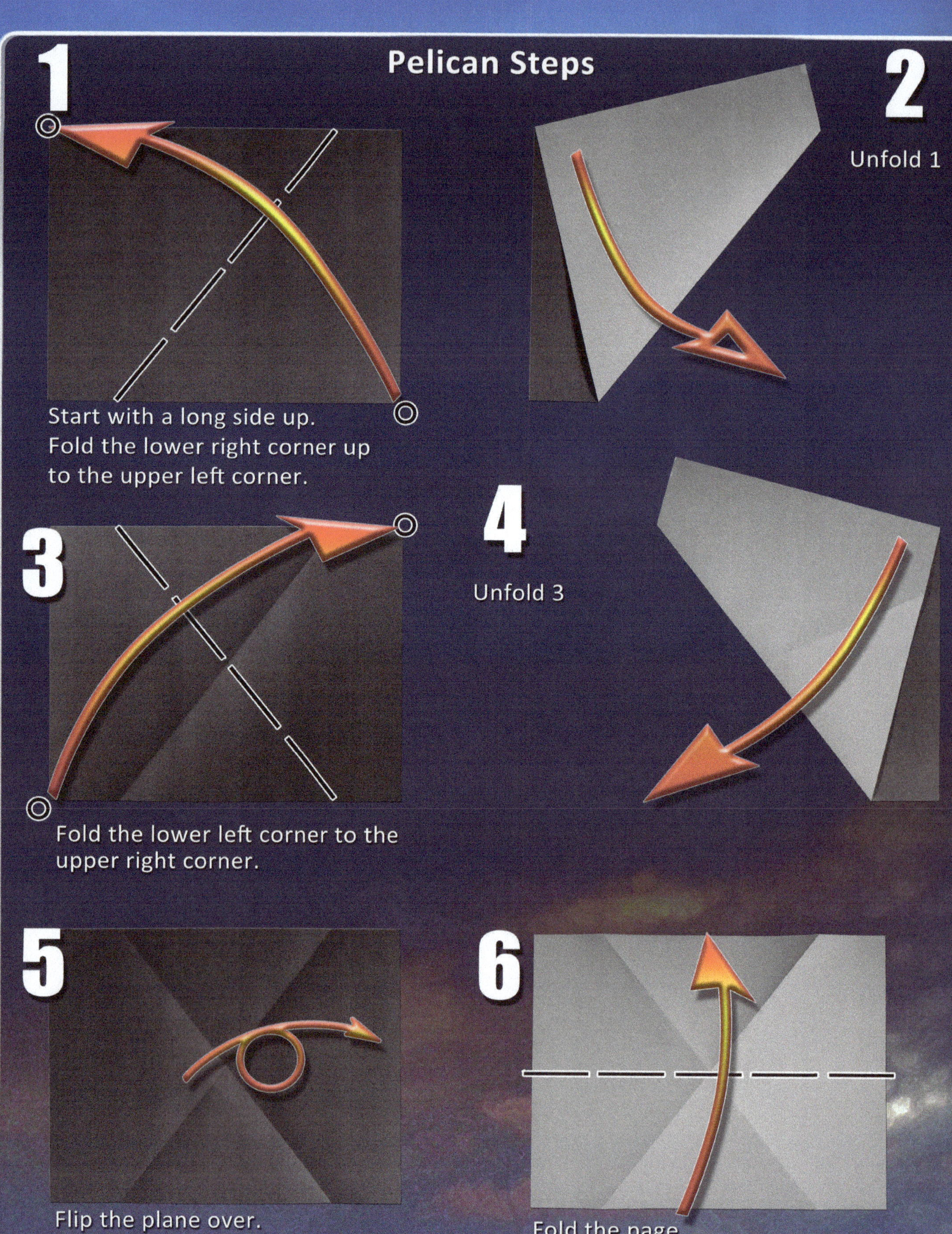

1 Start with a long side up. Fold the lower right corner up to the upper left corner.

2 Unfold 1

3 Fold the lower left corner to the upper right corner.

4 Unfold 3

5 Flip the plane over.

6 Fold the page in half.

Pelican Steps

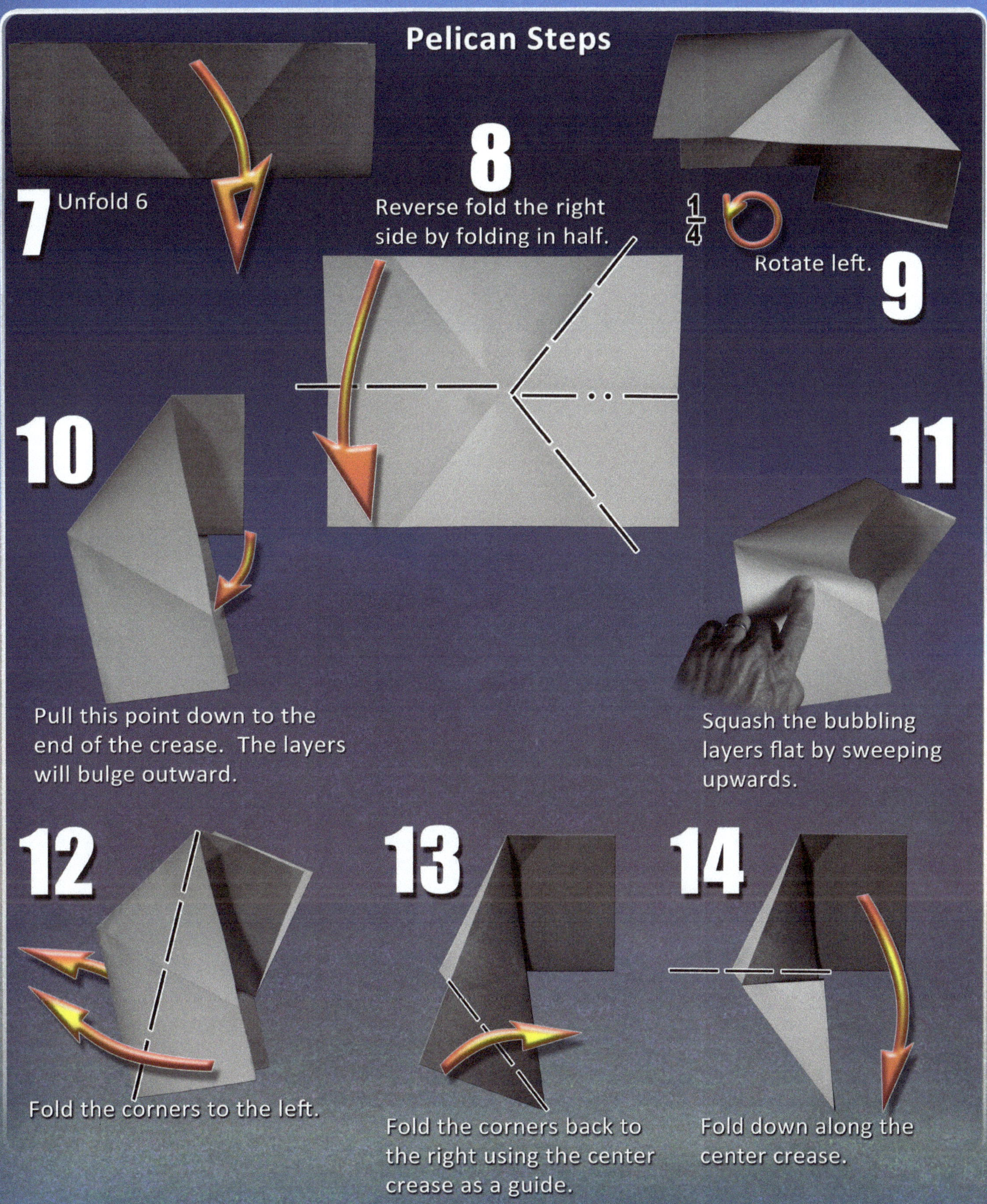

Pelican Steps

15

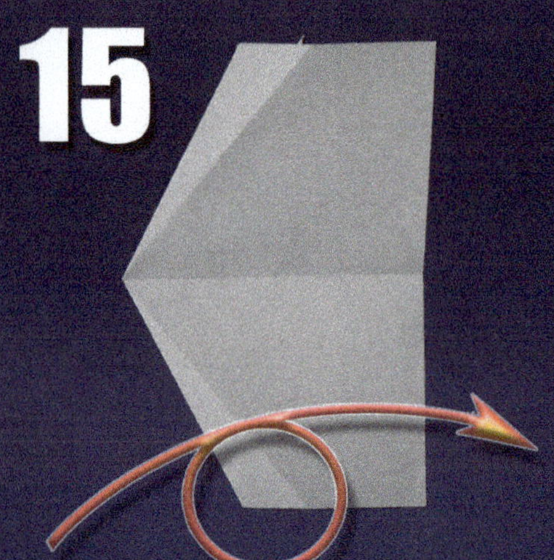

Flip the plane over.

16

Use the center crease as a guide.

17

Squash this point.

18

Fold the edges to the center. A petal fold is coming.

19

Unfold Step 18

20

Petal fold.

21

Flip the plane over.

22

Use the crease from 20 as a guide to make this crease.

23

Flip the plane over.

Pelican Steps

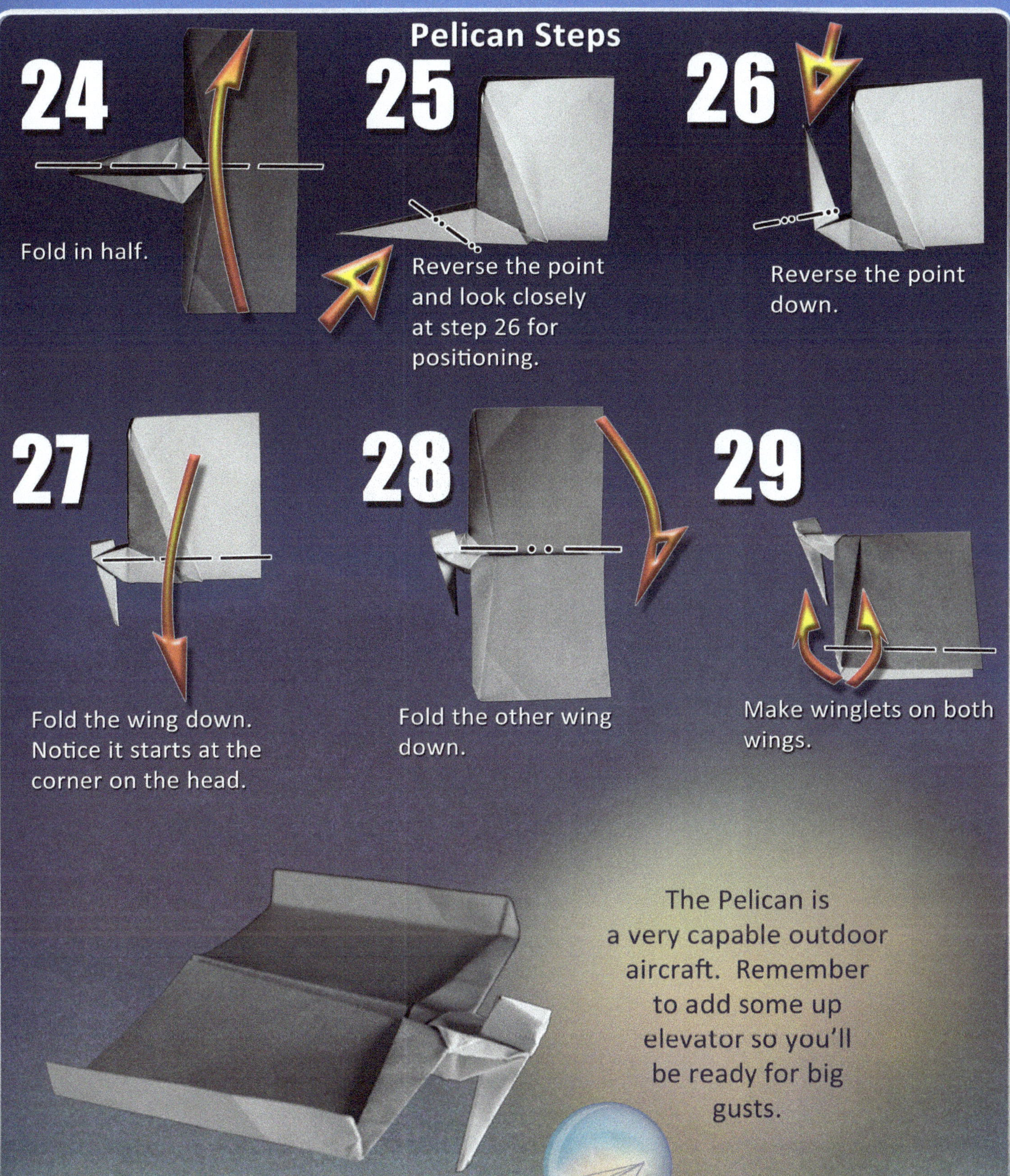

24 Fold in half.

25 Reverse the point and look closely at step 26 for positioning.

26 Reverse the point down.

27 Fold the wing down. Notice it starts at the corner on the head.

28 Fold the other wing down.

29 Make winglets on both wings.

The Pelican is a very capable outdoor aircraft. Remember to add some up elevator so you'll be ready for big gusts.

STARSHIP SHUTTLE

Glider

The Starship Shuttle was modeled after the shuttle craft on on a '60s tv series that spawned movies and many other series. The movies alone are now in their third cast line-up.

It can be finicky to adjust, but once you get the hang of it, the Starship Shuttle is very aerobatic. Loops and barrel rolls are easily executed.

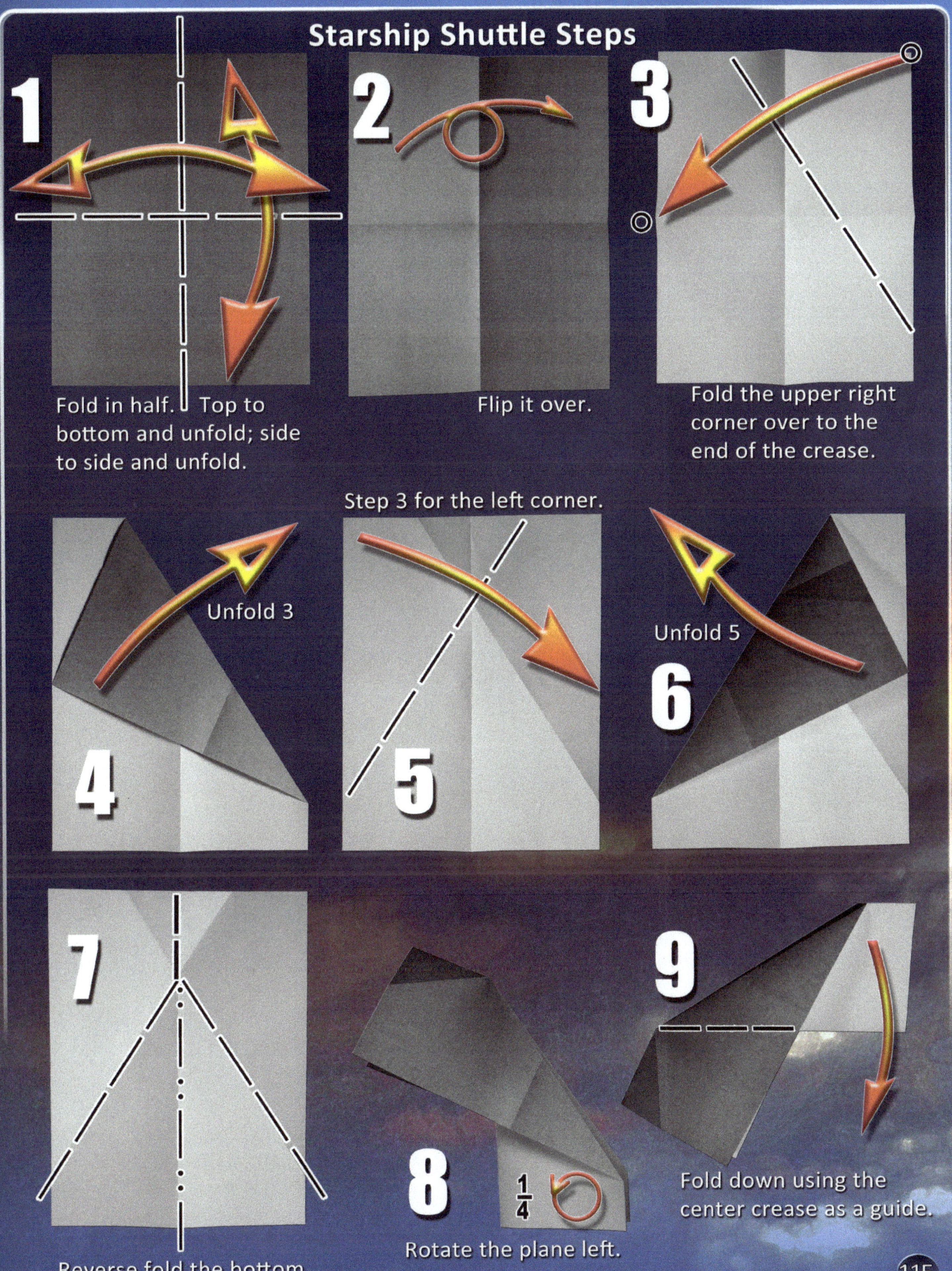

Starship Shuttle Steps
1
2
3
Fold in half. Top to bottom and unfold; side to side and unfold.
Flip it over.
Fold the upper right corner over to the end of the crease.
Step 3 for the left corner.
Unfold 3
4
5
Unfold 5
6
7
8
1/4
Rotate the plane left.
9
Fold down using the center crease as a guide.
Reverse fold the bottom.
115

Starship Shuttle Steps

10

Flip the plane over.

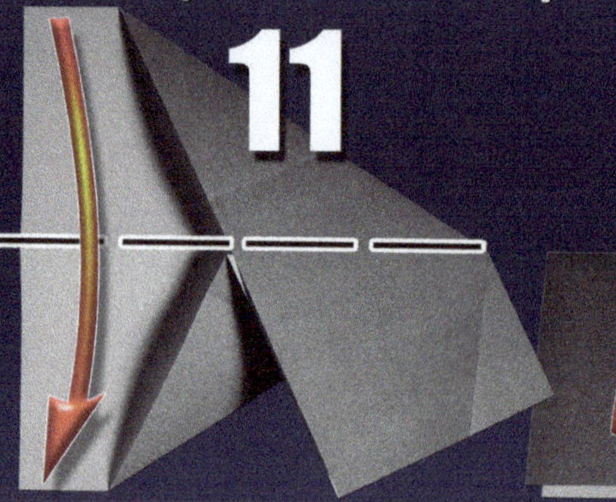

11

Fold the top half down using the center crease as a guide.

12

Open up the halves and squash the right side.

12a

Step 12 half way done, just before the completed squash.

12b

Complete the squash by bringng the point to the center.

13

Lift the top layer of the corner. Form two creases. One uses the top layer raw edge. The other goes corner to corner.

13a

One crease made.

14

Repeat 13 for this side.

15

Fold the corner to the point where the layers meet.

Starship Shuttle Steps

16

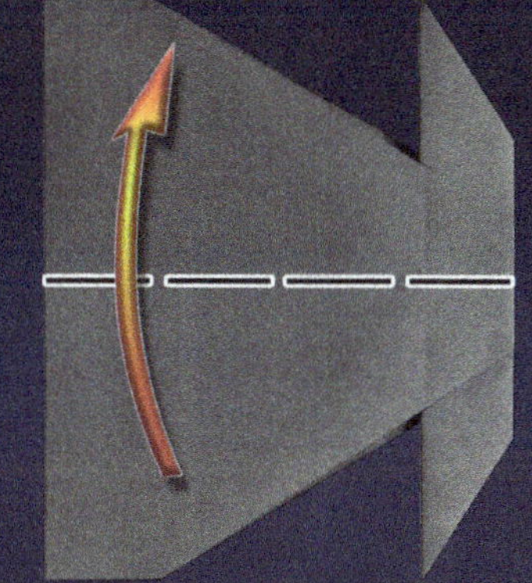

Flip the plane over.

17

Fold the plane in half.

18

The wing crease slopes gently up toward the tail.

19

Open up the wings so the plane is resting on the tops of the wings.

20

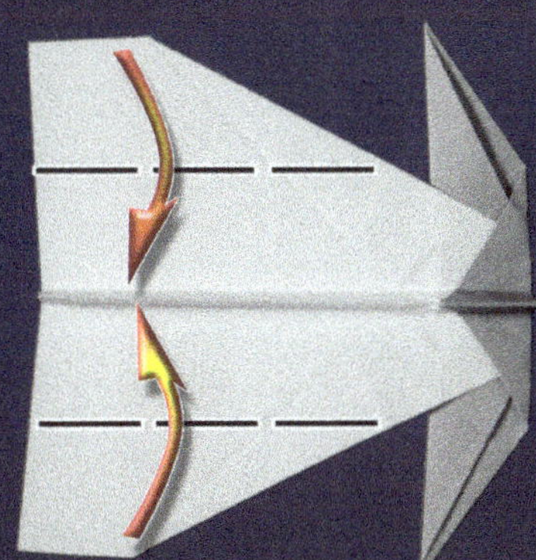

Fold the raw edges to the center.

21

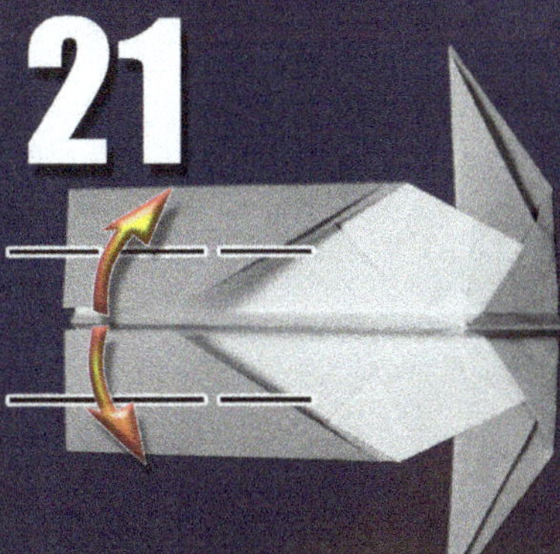

Fold the raw edges back to the creased edge.

22

Fold the wing back over.

23

Fold the flaps up.

24

Time to connect the flaps.

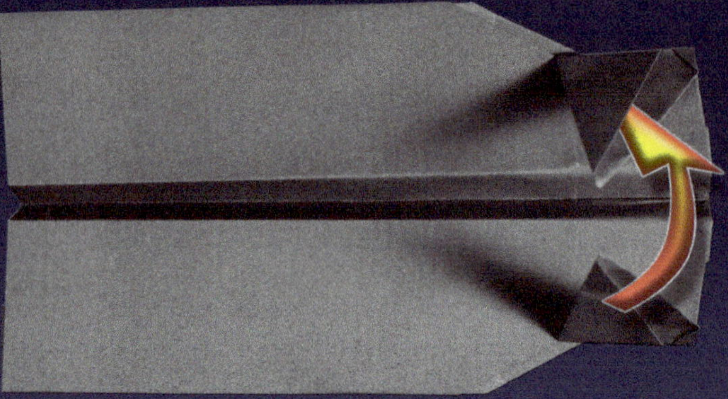

25 Put one flap inside the other like this.

26 Make a crease through the center of the overlap. Then lock the flaps together with a small reverse fold as shown.

Open up the wings and make an upward bend at the rear corners. It needs that much up elevator to keep flying!

RING THING

Glider

The ring in front technically makes the Ring Thing a canard design.

It's a bit clunky in the air, but that's not why you're folding this plane. It's the challenge of doing the nearly impossible.

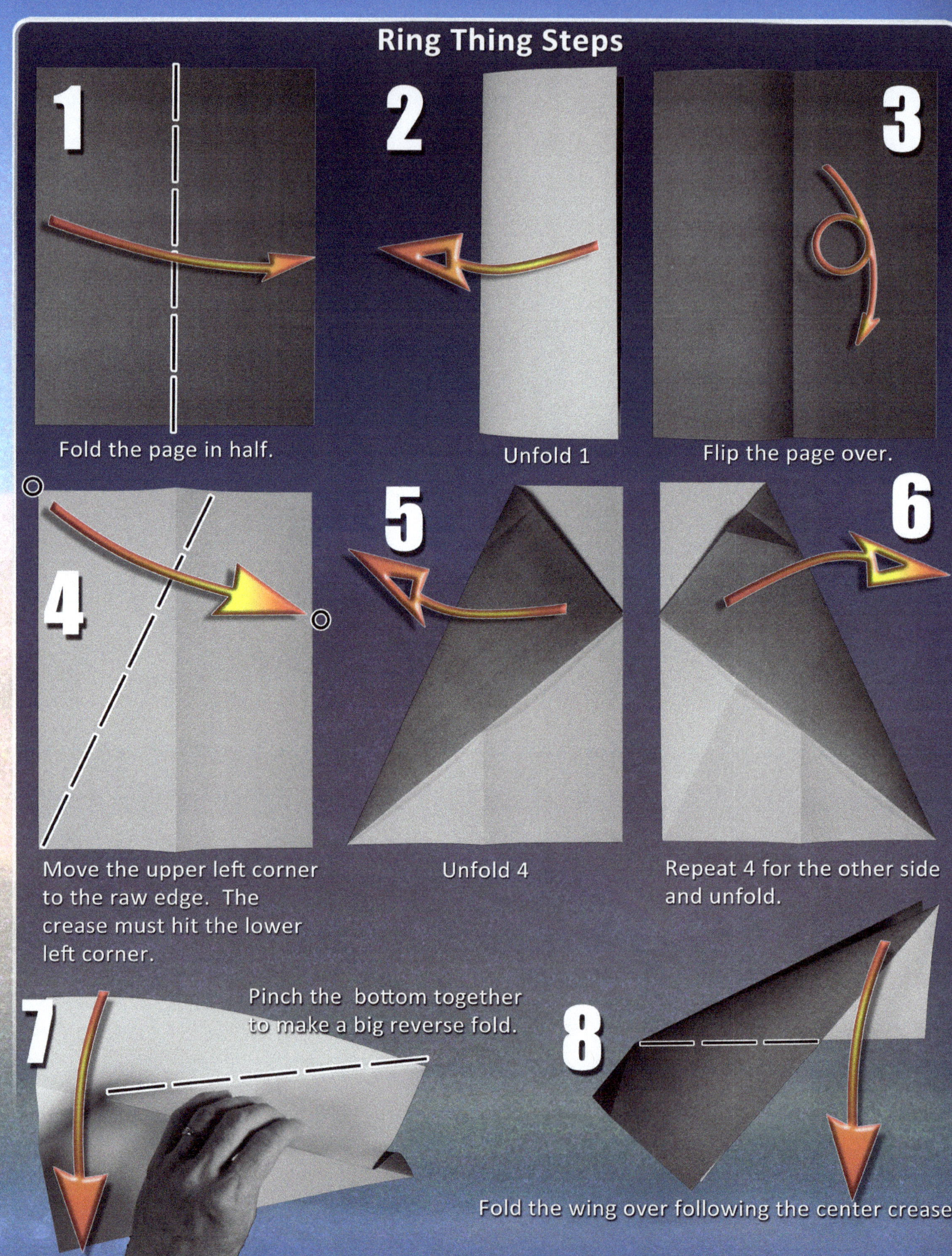
Ring Thing Steps
1
Fold the page in half.
2
Unfold 1
3
Flip the page over.
4
Move the upper left corner to the raw edge. The crease must hit the lower left corner.
5
Unfold 4
6
Repeat 4 for the other side and unfold.
7
Pinch the bottom together to make a big reverse fold.
8
Fold the wing over following the center crease.
120

Ring Thing Steps

9

Flip the plane over.

10

Fold the wing down.

11

Open up the wings and squash the right side.

11a

Wings are opened. Squash this part.

12

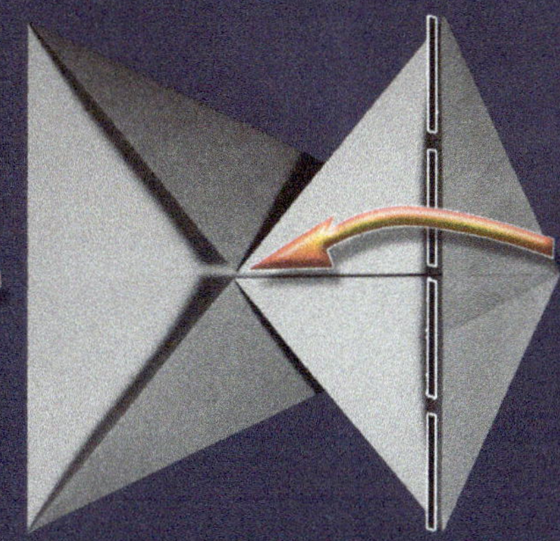

Fold the top triangle over.

13

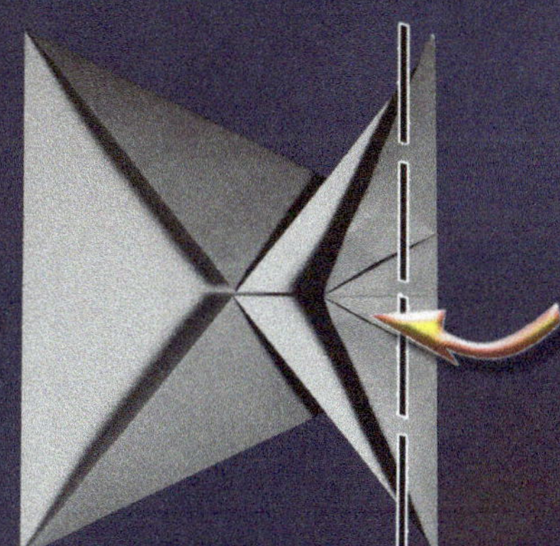

Fold over 1/3 of the triangle.

14

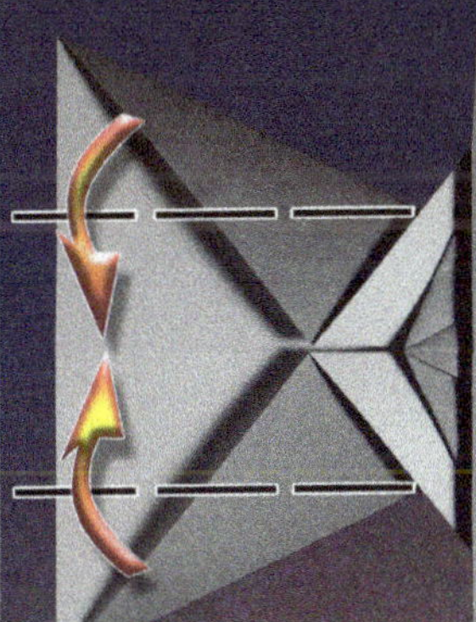

Fold the corners past the center using the top layer as a guide.

15

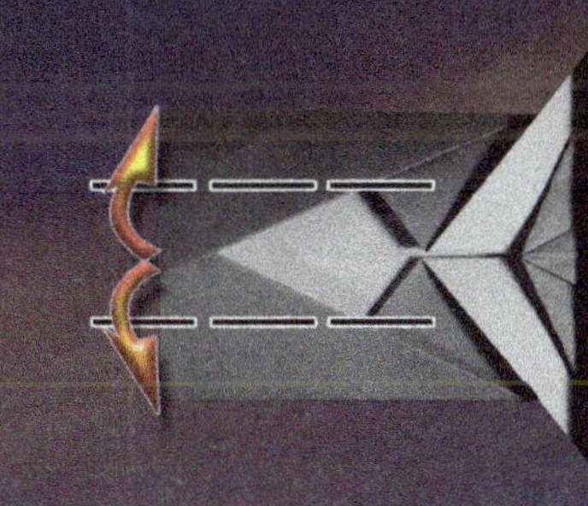

Fold the corners back to the creased edges.

16

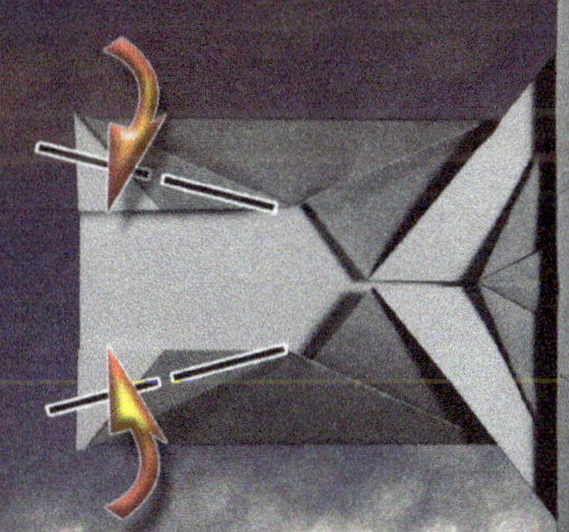

This step is shown a bit larger so you can easily see the small creased edge is lining up with the other creased edge.

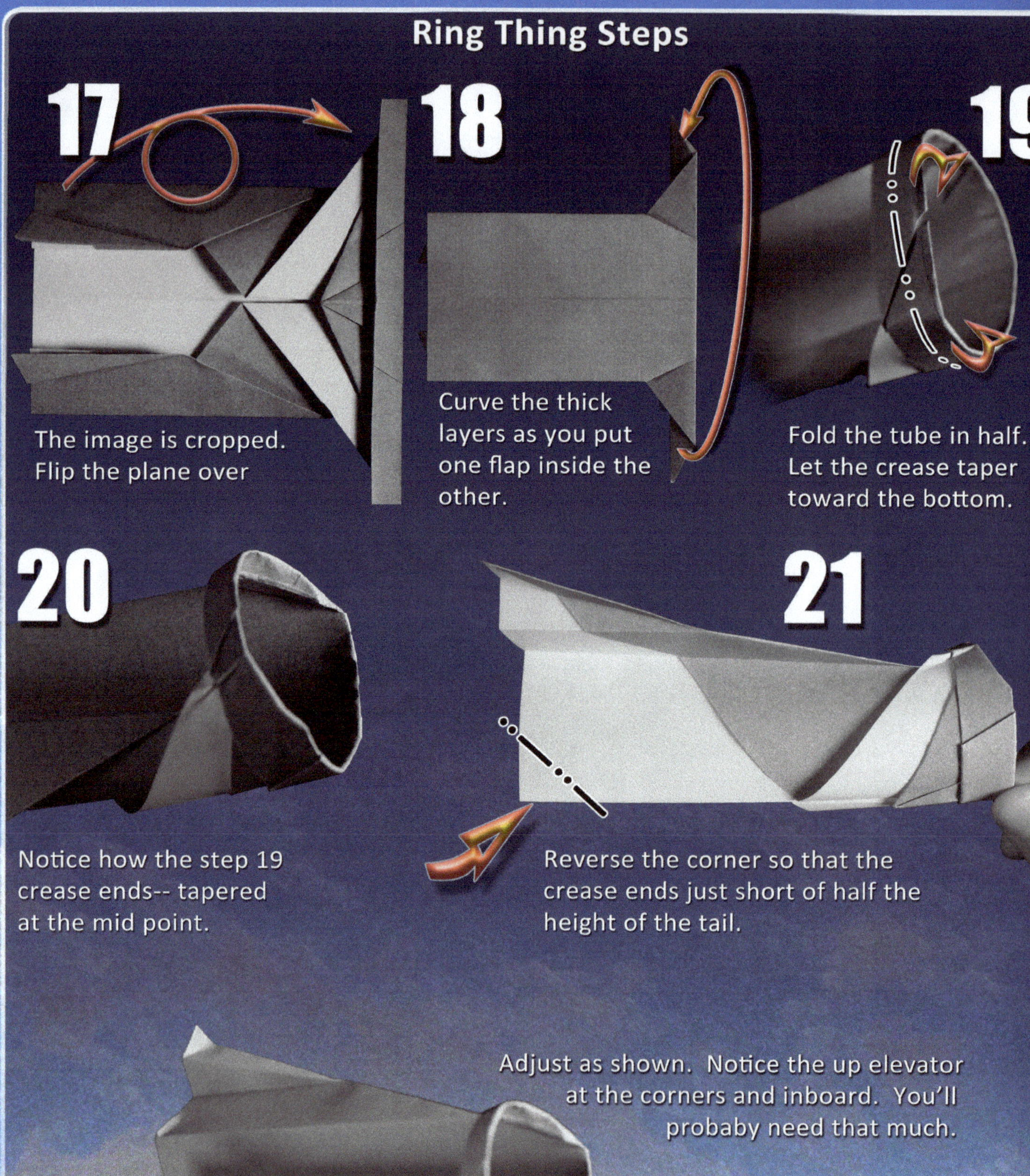

Ring Thing Steps

17

18

19

The image is cropped.
Flip the plane over

Curve the thick
layers as you put
one flap inside the
other.

Fold the tube in half.
Let the crease taper
toward the bottom.

20

21

Notice how the step 19
crease ends-- tapered
at the mid point.

Reverse the corner so that the
crease ends just short of half the
height of the tail.

Adjust as shown. Notice the up elevator
at the corners and inboard. You'll
probaby need that much.

SUPER CANARD

Glider

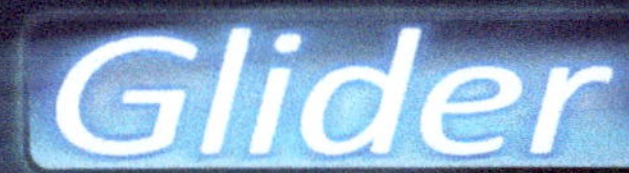

The Super Canard is a great paper airplane. The small wing in front can be adjusted to make the plane stall resistant.

Path

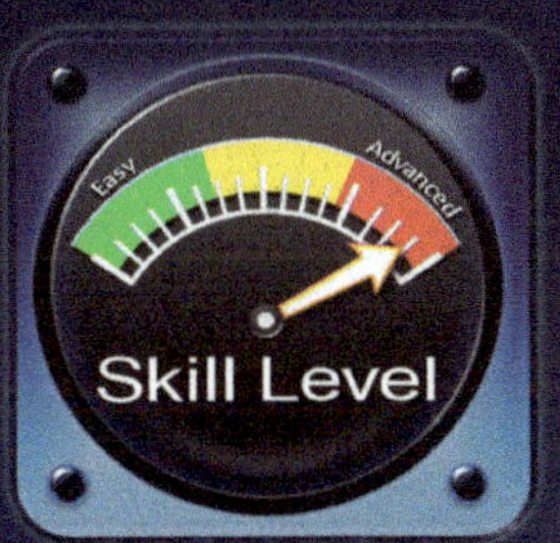

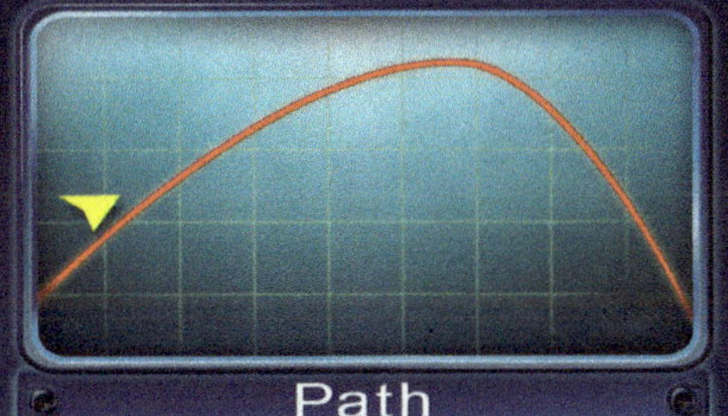

The folding is a challenge, and the result is worth the trouble. There's a lot to keep track of during testing and adjusting. Take your time.

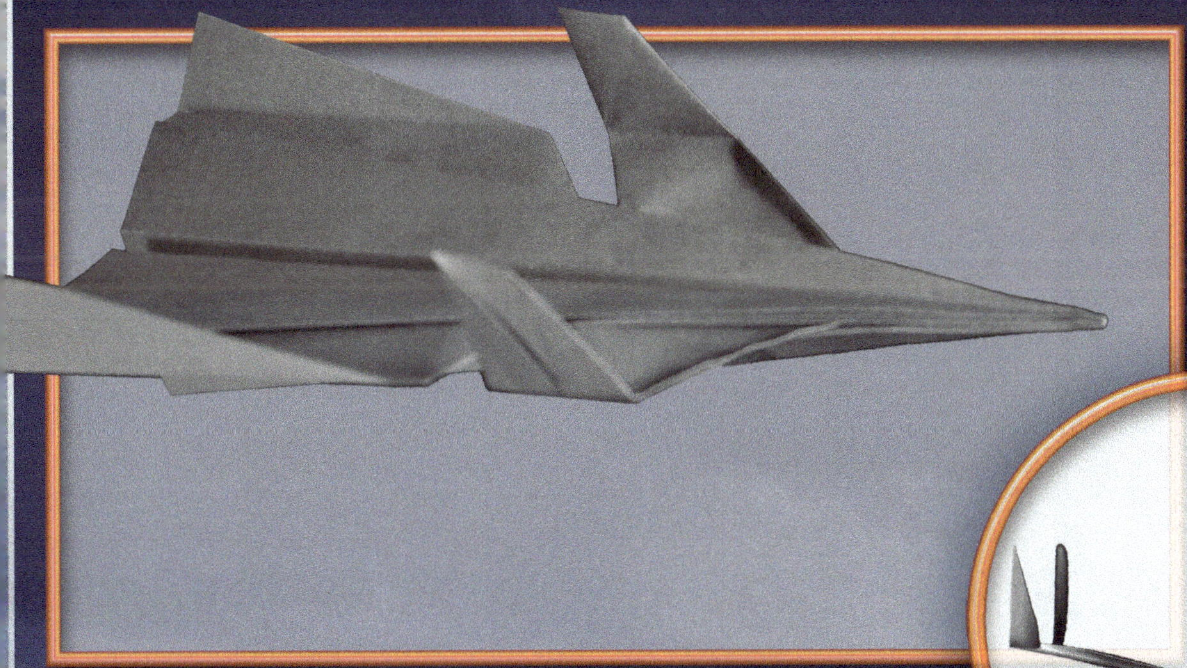

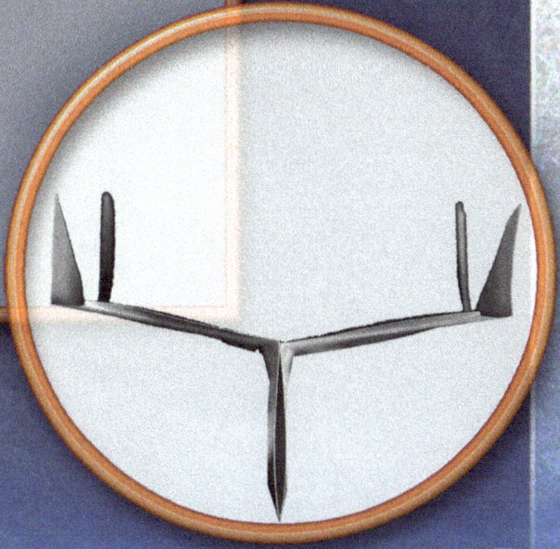

Super Canard Steps

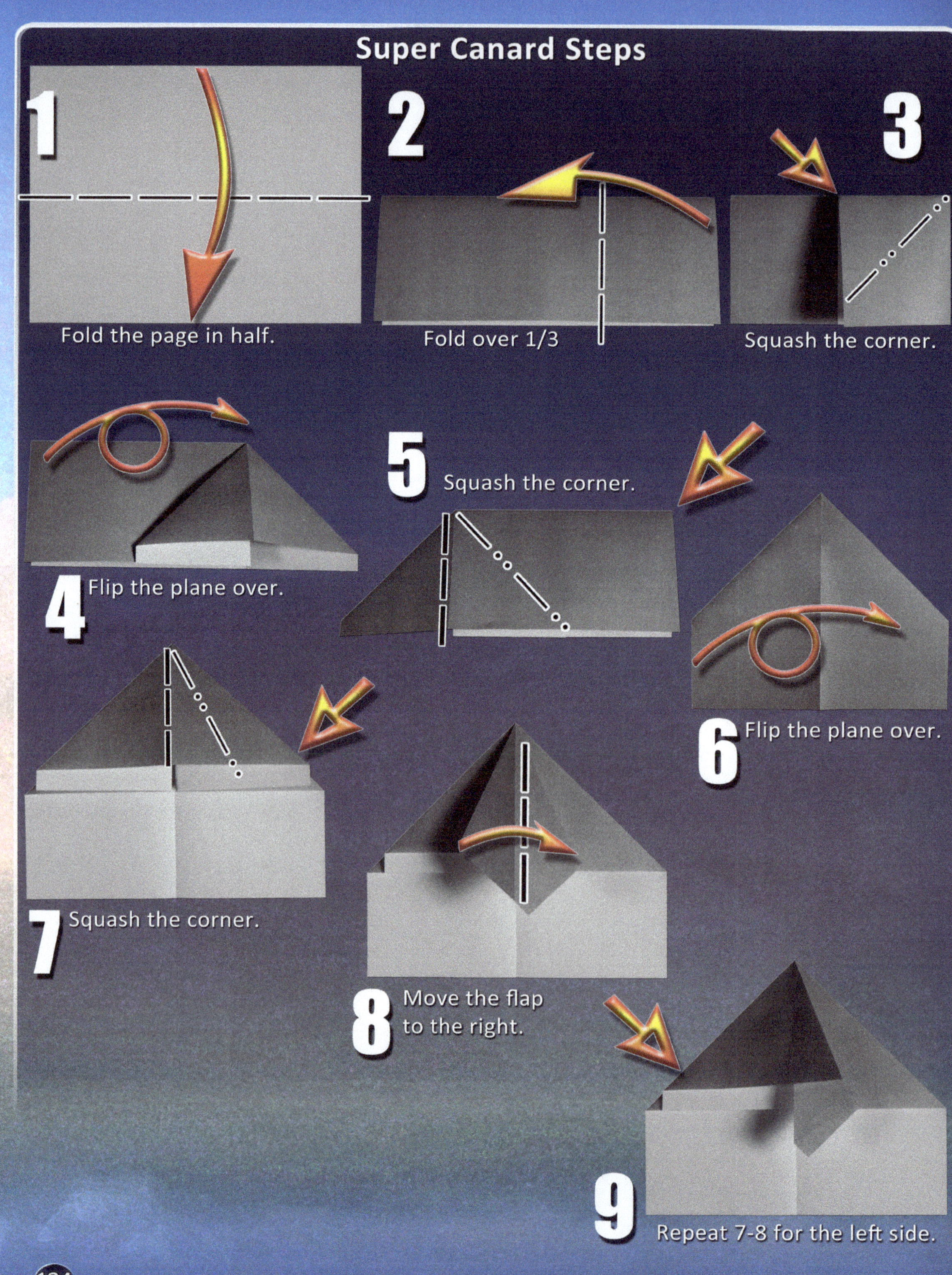

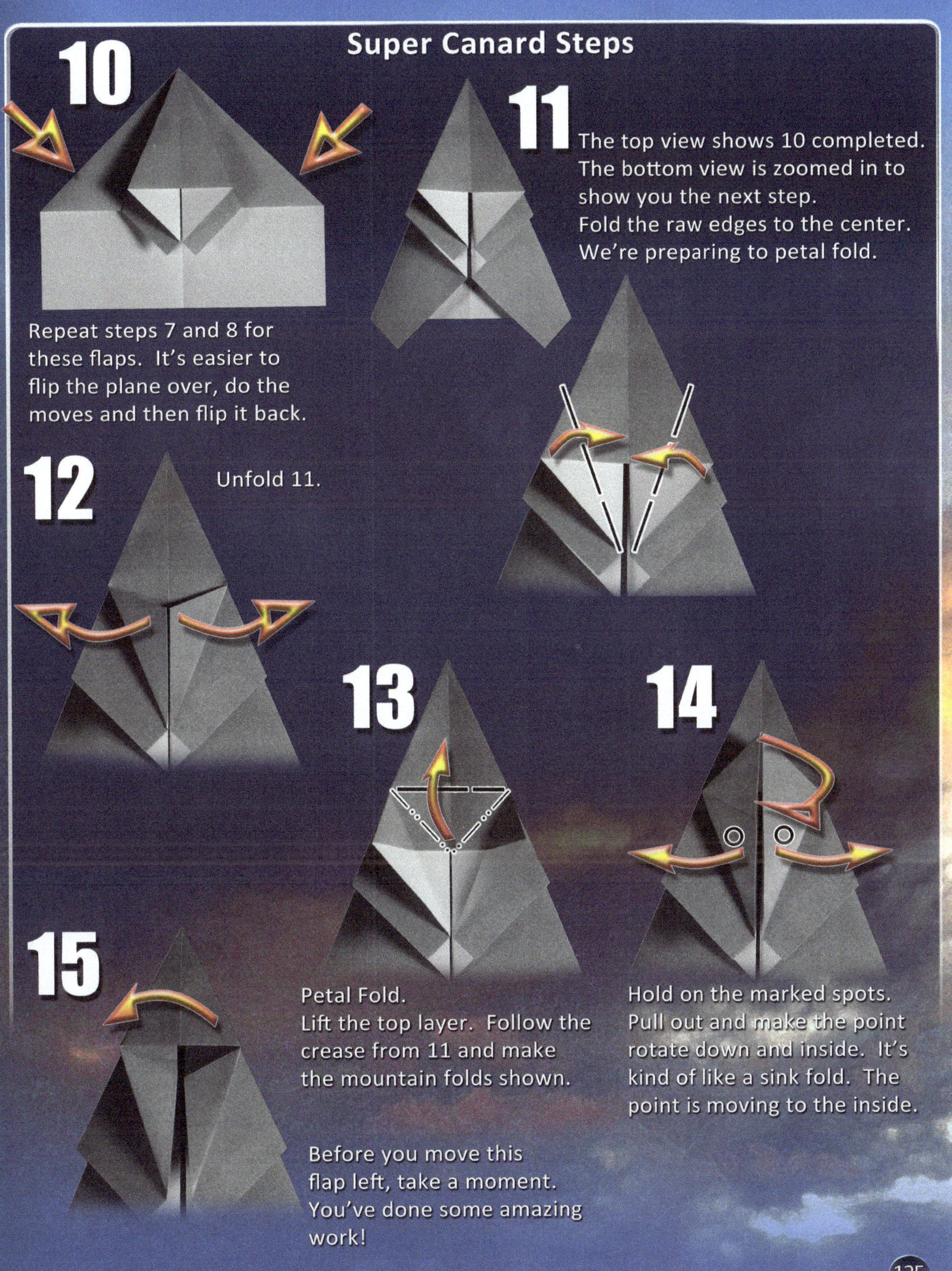
Super Canard Steps

10

11
The top view shows 10 completed.
The bottom view is zoomed in to
show you the next step.
Fold the raw edges to the center.
We're preparing to petal fold.

Repeat steps 7 and 8 for
these flaps. It's easier to
flip the plane over, do the
moves and then flip it back.

12

Unfold 11.

13

14

15

Petal Fold.
Lift the top layer. Follow the
crease from 11 and make
the mountain folds shown.

Hold on the marked spots.
Pull out and make the point
rotate down and inside. It's
kind of like a sink fold. The
point is moving to the inside.

Before you move this
flap left, take a moment.
You've done some amazing
work!

125

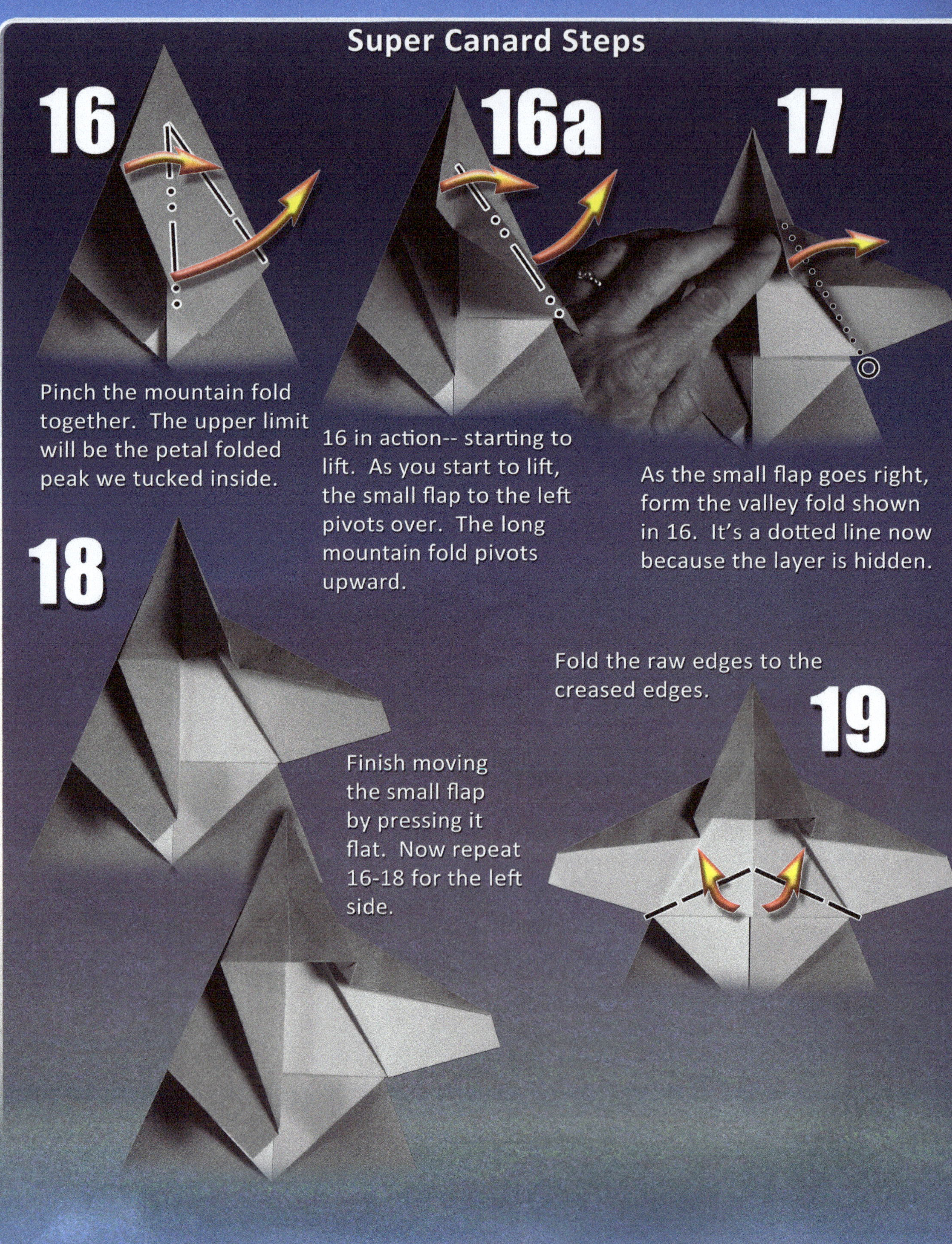

16

Pinch the mountain fold together. The upper limit will be the petal folded peak we tucked inside.

16a

16 in action-- starting to lift. As you start to lift, the small flap to the left pivots over. The long mountain fold pivots upward.

17

As the small flap goes right, form the valley fold shown in 16. It's a dotted line now because the layer is hidden.

18

Finish moving the small flap by pressing it flat. Now repeat 16-18 for the left side.

Fold the raw edges to the creased edges.

19

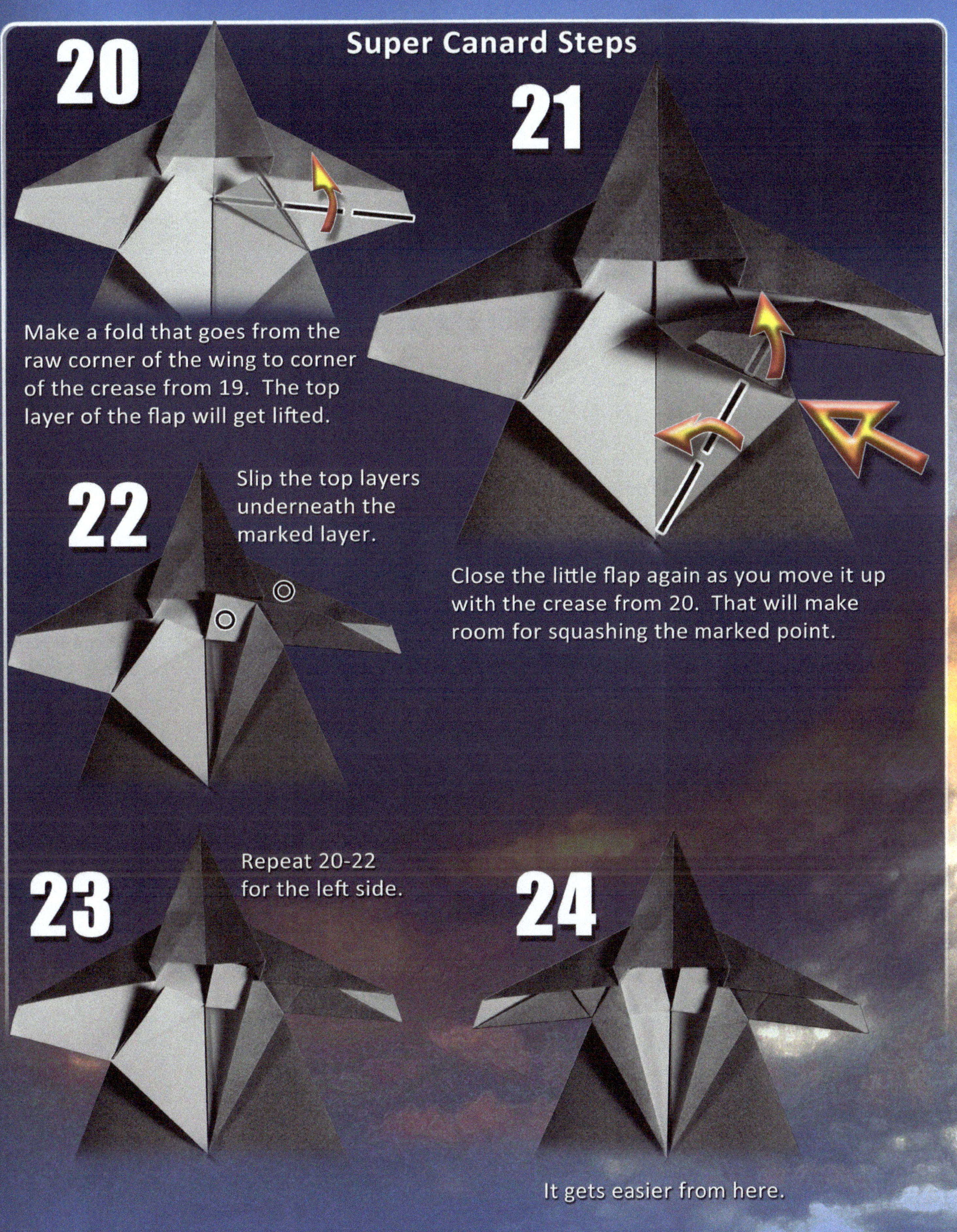

Super Canard Steps

20

21

Make a fold that goes from the raw corner of the wing to corner of the crease from 19. The top layer of the flap will get lifted.

22

Slip the top layers underneath the marked layer.

Close the little flap again as you move it up with the crease from 20. That will make room for squashing the marked point.

23

Repeat 20-22 for the left side.

24

It gets easier from here.

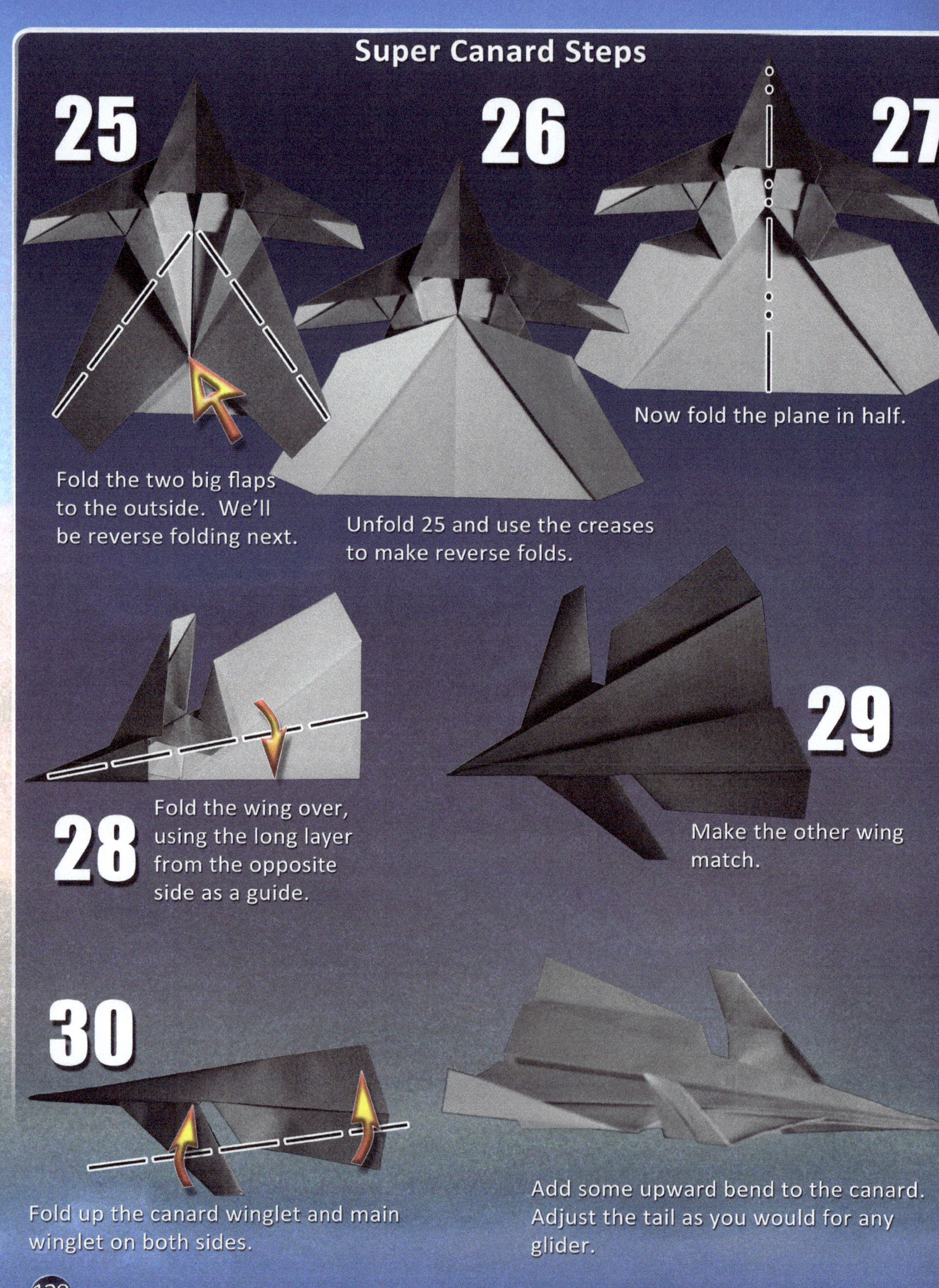

Super Canard Steps

SEAGULL

Designed by Ryan Naccarato, the winningest competitor in Red Bull Paper Wings history.

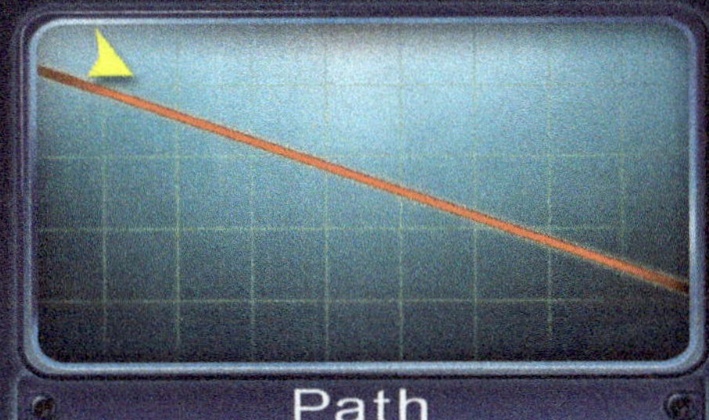

Simple to fold, and last in the book because this plane is used as the big finish in winning routines. Dozens, sandwiched between paper plates, get launched high up and float gently down.

So light and graceful, the Seagull is also a great Follow Foil!

Seagull Steps

1

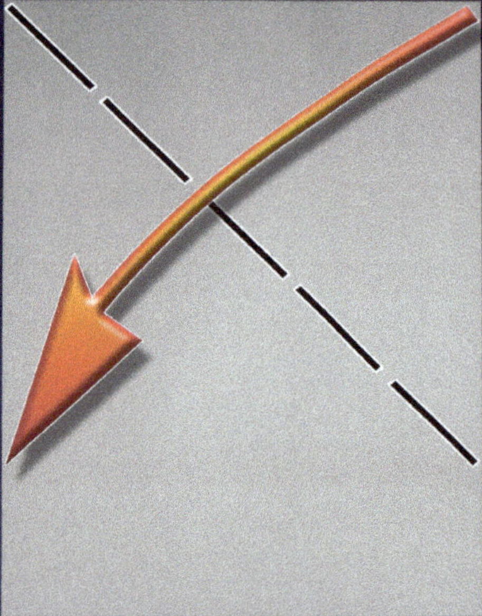

Make a diagonal fold. Line up the top with the left side.

2

Unfold step 1

3

Cut just barely above the crease.

4

The triangle is rotated to the right. Fold it in half.

Seagull Steps

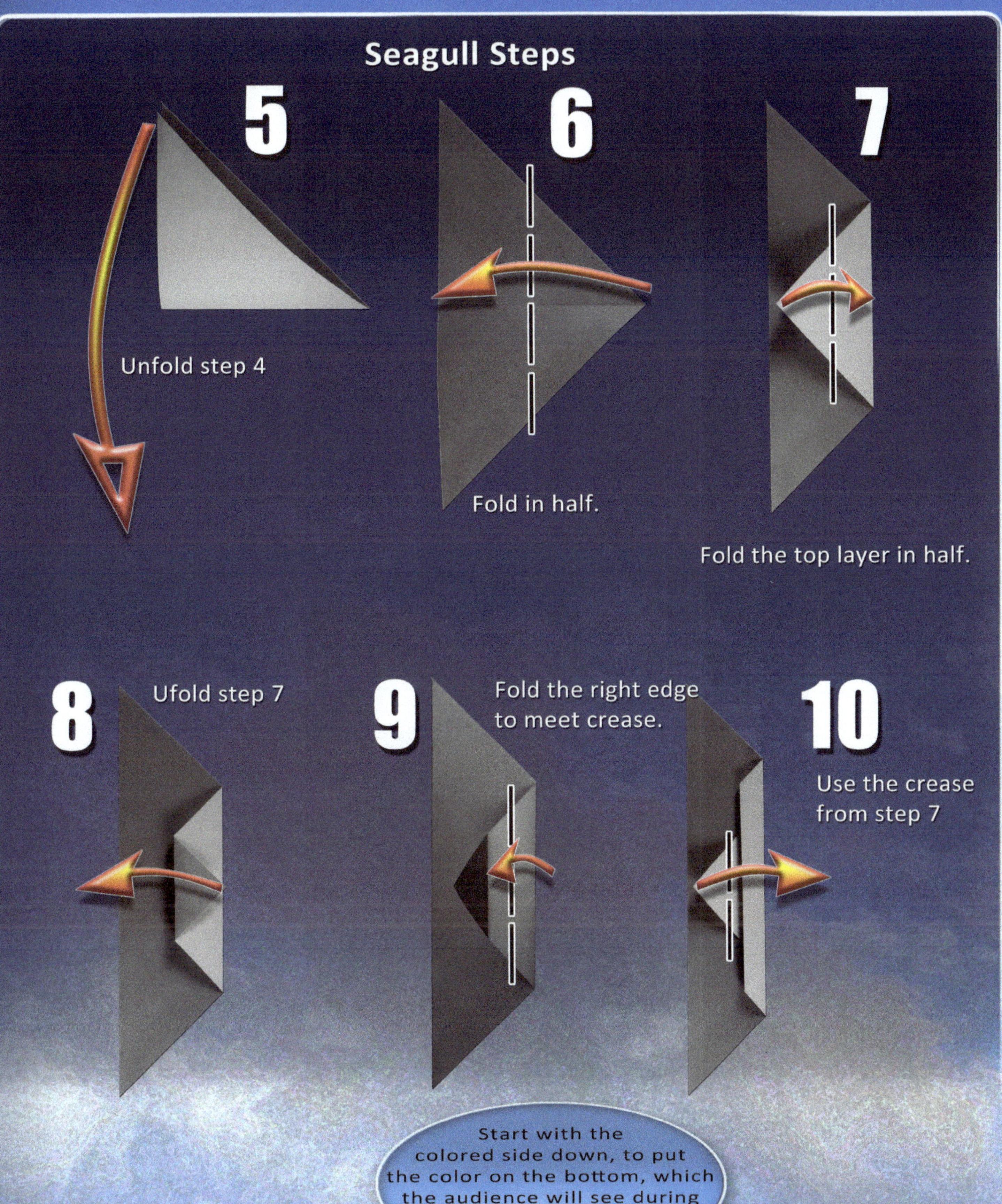

Seagull Steps

11

12

12b

90°

Notice how the crease hits the corner at the right. Also, the idea is to create a square corner when step 12 is done.

Follow the crease.

This step is zoomed in a little. Fold the corner down. Look at the next step for details.

13

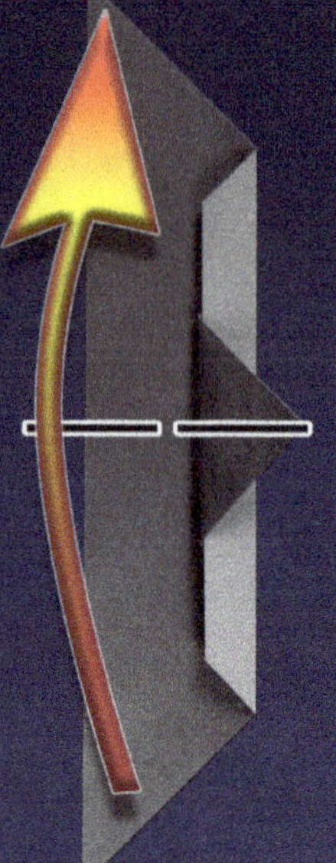

Set the wing angles and drop from a good height.

Fold the other side down. Then open up the plane.

The Flight Stuff

Know your aircraft parts.
Notice all the control surfaces are at the trailing edge.

If the plane turns right, bend the rudder edges right (away from us here)
If the plane turns left, bend the rudder edges left (at us in this picture)
If the plane dives, bend the elevator edges up.
If the plane climbs too much, bend the elevator edges down.

Before making any adjustment, make sure all your creases are sharp, one side of the plane matches the other, and the trailing edges are straight.

Throwing

Throwing and adjusting work together. Tiny adjustments to rudder and elevator surfaces make a huge difference in the flight. Getting the throw right makes the most of all your careful work.

First, get a grip. Hold your plane where most of the layers meet. The Thickest part of the plane is close the the center of gravity (CG). Pushing against the CG helps the plane go straight. We'll talk more about the CG later.

It's best to get the plane gliding and flying straight first, and then throw it harder. The first few throws should assist the plane into the air.

Sharp creases, symmetry (the left side matches the right side) and crisp, clean trailing edges are the starting points.

Try to launch with the wings level. Use a smooth motion to put less stress on the wings. After a couple of throws, you'll know about how fast the plane glides. Try to match that speed with your launch as you're dialing in a straight, smooth glide.

Basic Forces

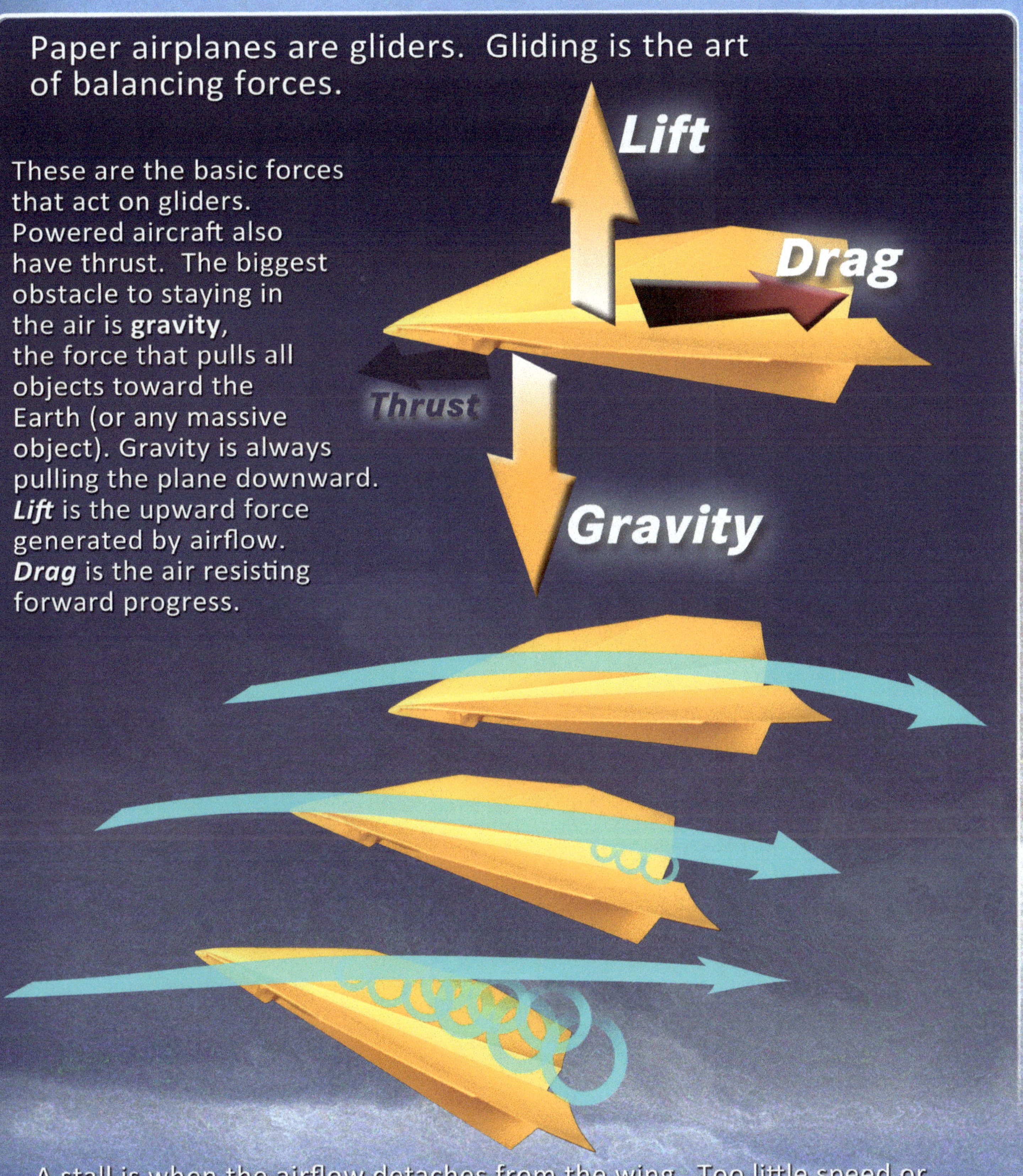

Paper airplanes are gliders. Gliding is the art of balancing forces.

These are the basic forces that act on gliders. Powered aircraft also have thrust. The biggest obstacle to staying in the air is **gravity**, the force that pulls all objects toward the Earth (or any massive object). Gravity is always pulling the plane downward. *Lift* is the upward force generated by airflow. *Drag* is the air resisting forward progress.

A stall is when the airflow detaches from the wing. Too little speed or angling the wing too sharply into the direction of airflow will cause the air to stop following the the shape of the wing.

Lift is the result of airflow around a properly shaped wing. A combinaton of a lower pressure on the upper surface, and the air directed downward off the trailing edge, creates an upward force on the wing. We call that lift. With enough airflow, we can generate enough lift to overcome gravity.

Drag is the result of airflow interacting with the shape and surface of the aircraft; paper in this case. Drag is constantly slowing the plane's progress.

Gravity is the force that pulls all objects toward the Earth (or any massive object). Gravity creates a downward force on our airplane.

Balancing these three forces keeps our airplanes flying. While it's obvious that lift overcomes gravity at some airspeed, it's not obvious how our airplanes will win against drag. Drag is constantly slowing the plane. Left un-checked, drag would slow the wing to the point of stalling. Powered planes simply hit the throttle, creating thrust. Gliders have another solution.

Center of Gravity is the secret weapon. Every object has a center of gravity. It's the point around which the object rotates. Whether it's a planet or an asteroid tumbling through space, everything rotates around its center of gravity (CG). Physicists like to say objects *translate* through their CG. That's a fancy way of saying "follow the path of". We plot the path of an object through its CG.

Trading Height for Speed

How the Center Of Gravity helps the plane fly

Drag is slowing the plane. Lift and gravity are the only tools available. The solution: trade a little height for some speed. We use gravity to gain speed. In a well made plane, we add just enough speed to make up for the drag.

To gain speed, we need to point the nose of the plane toward the Earth. The plane is already flying, so now what?. Lift also has an average point that we call *Center of Lift*. Center of lift is point where lift appears to be acting, very similar to CG. The plane rotates around the CG, so you can think of CG like the center of a teeter totter.

If the center of gravity is in front of the center of lift, the back of the plane rotates up. The nose of the plane is on the other side of the teeter totter, so it rotates down. This is the common way paper airplanes gain speed to overcome drag.

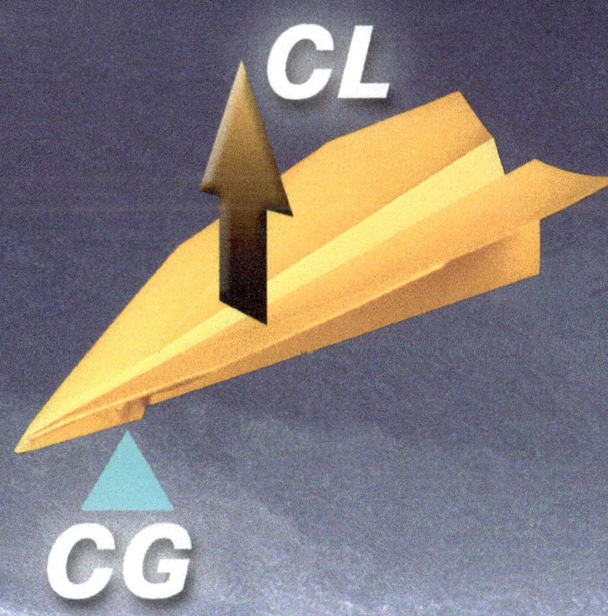

The plane is now headed on a downward path. It's gaining speed. Left alone, it will crash. What's your next move?

Up elevator is bending the trailing edge a little upward.
The air hits that bend and bounces upward.
The bounce force pushes the tail down.
The plane rotates around
the center of gravity.
The nose of the
plane rotates up.

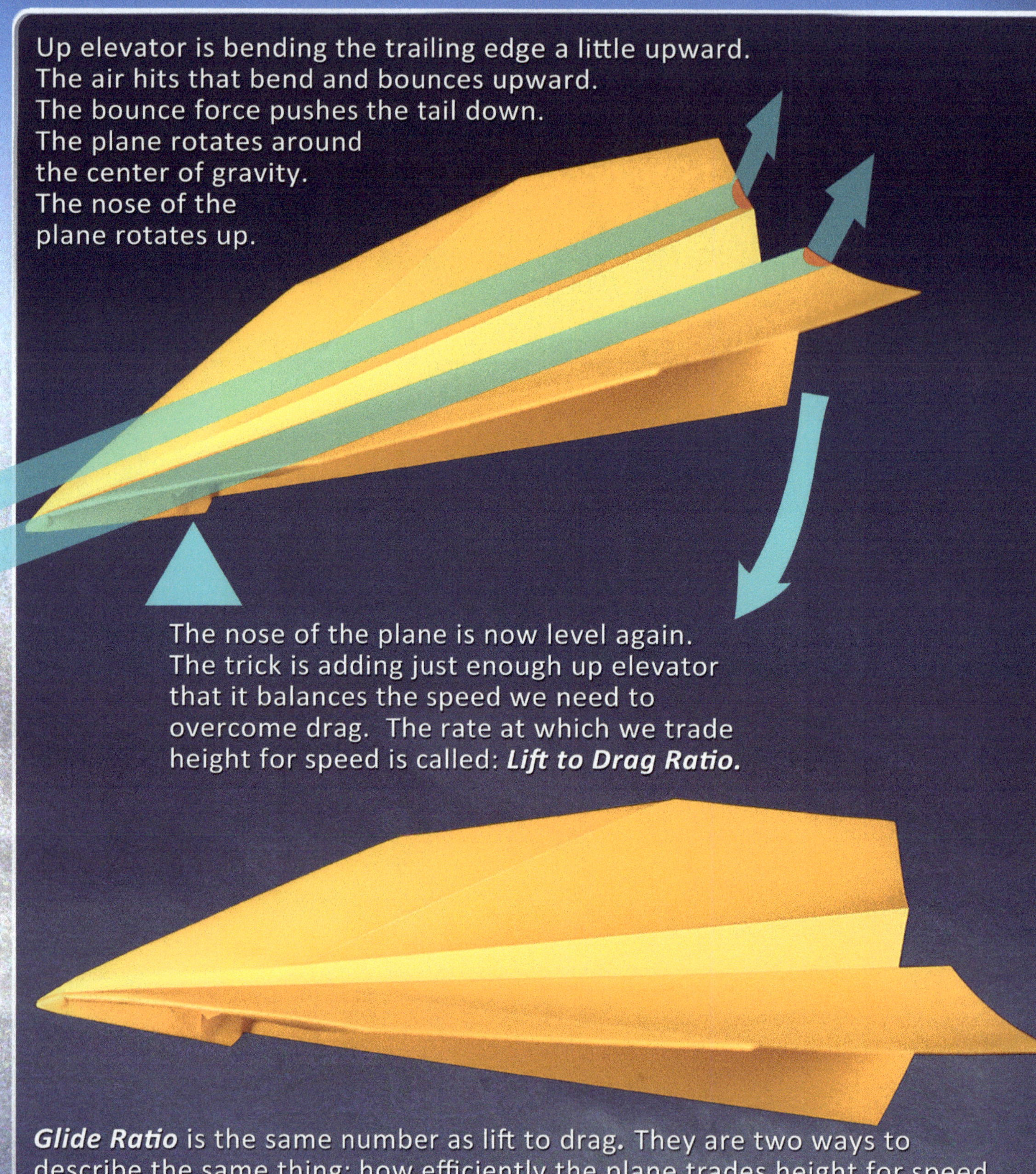

The nose of the plane is now level again.
The trick is adding just enough up elevator
that it balances the speed we need to
overcome drag. The rate at which we trade
height for speed is called: *Lift to Drag Ratio.*

Glide Ratio is the same number as lift to drag. They are two ways to
describe the same thing: how efficiently the plane trades height for speed.
Glide Ratio expresses the number of feet of forward progress for each foot
of height lost; for example 6 to 1.

All adjustments for turning left, right, and climbing work the same way as
up elevator: air bounces one way, causing the plane to rotate the other.

Steering in 3D

The delicate balance between drag, center of gravity, center of lift, and speed keeps the plane gliding smoothly. Changing the shape of the wing changes where the CL locates. Layering paper moves the CG. The CG forward of the CL allows the plane to gain speed. The right amount of speed reflects enough air off the elevator to keep the nose level and fend off drag.

Roll

We've covered the basics of how a plane rotates around the center of gravity to climb, dive, and turn left or right. There's another important direction: rolling. Bending the rear corners in opposite directions will make the plane barrel roll. Of course constantly rolling isn't really helpful for most flights.

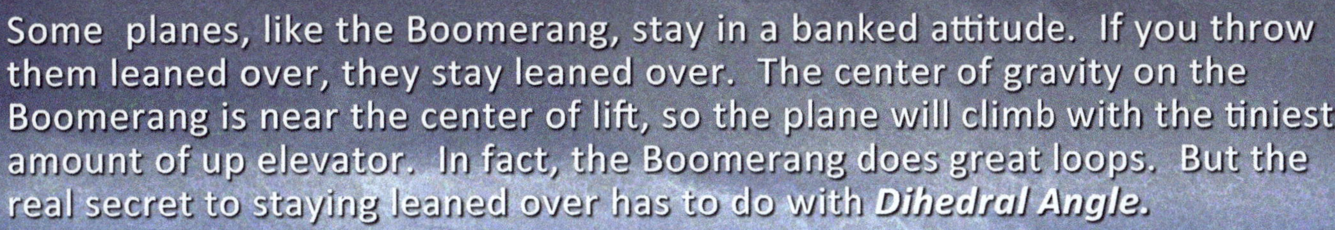

Some planes, like the Boomerang, stay in a banked attitude. If you throw them leaned over, they stay leaned over. The center of gravity on the Boomerang is near the center of lift, so the plane will climb with the tiniest amount of up elevator. In fact, the Boomerang does great loops. But the real secret to staying leaned over has to do with *Dihedral Angle.*

Dihedral Angle is simply the angle the wings are attached to the plane. Most paper airplanes (good ones) have positive dihedral, which is a fancy way of saying the wings slope upward as they leave the fuselage (body) of the plane.

Postive Dihedral puts the lifting surface above the center of gravity. So, if the plane gets rocked to one side, sort of like a pendulum, the plane rocks back to neutral. This is sometimes called ***Dead Stick Stability.***

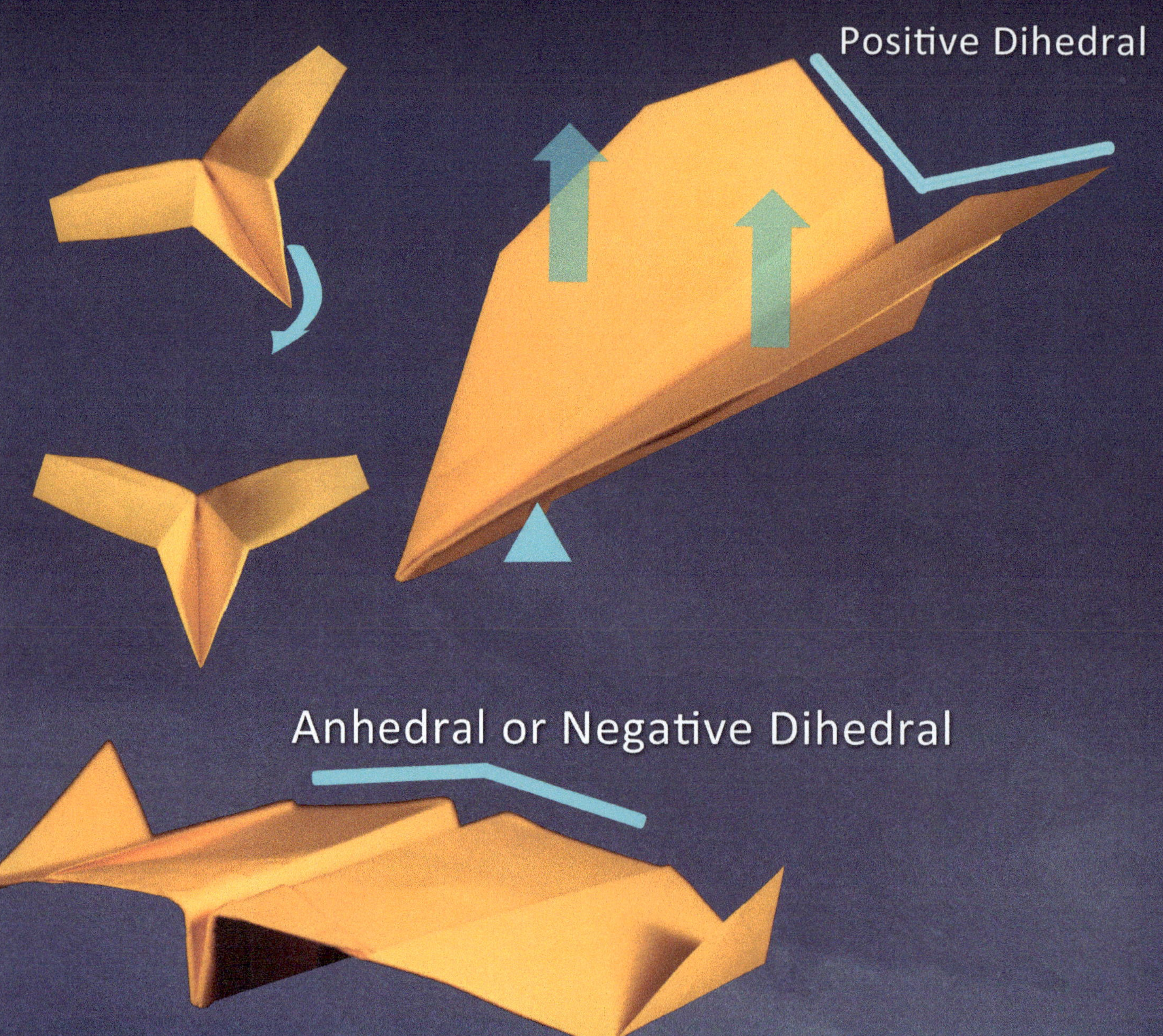

Take a close look at the Boomerang. The wings are drooping. That's called anhedral angle or negative dihedral. The plane doesn't rock back to neutral. If you throw it leaned over, it stays leaned over. Remember that climb? A climb, with the plane leaned to one side, makes a circle. That's why the Boomerang can circle right or left and loop.

That's just how we roll... Kutta-Joukowski wise

The Tube is that rare paper airplane that uses rapid rolling to create lift. It's also interesting to note that the CG is not on the plane. Remember, the center of rotation is the center of gravity, which in this case, is in thin air!

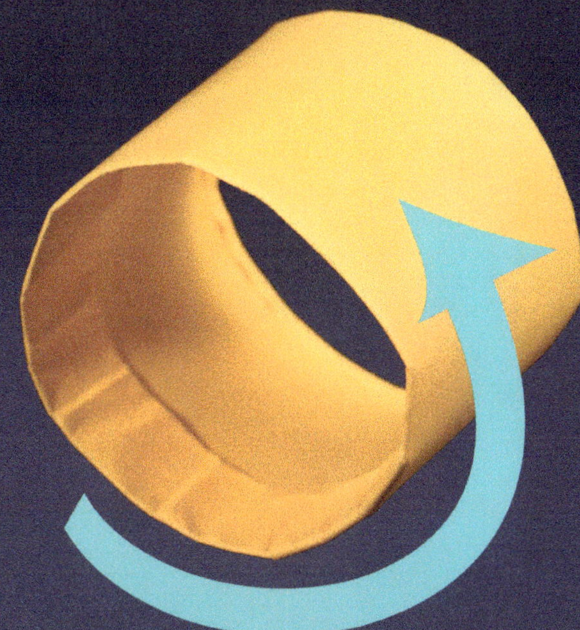

Just like a bicycle wheel, the faster The Tube spins, the easier it holds its course. Stability from spinning motion is called *angular momentum*. If you were thinking about the term, centrifugal force or gyroscopic force, you're not wrong.

It's simply more precise to say angular momentum. If you're throwing The Tube with enough spin, and hard enough, you'll notice (if you're right-handed) The Tube will curve gently to the right. In fact, a fun thing to practice is throwing slightly down and to the left of your target, and getting The Tube to curve up and right for a direct hit. For lefties, aim down and to the right. Why does this work? Glad you asked.

The Tube generates lift from the act of spinning. Spinning cylinders make a special kind of lift called *Kutta-Joukowski Lift.* In a spinning sphere (ball) it's called *Magnus Effect*. Magnus effect is why a curve ball in baseball works and why a golf ball hooks or slices. A layer of air gets trapped at the surface of the ball by the texture. The layer of air, spinning with the ball, becomes an air stream that interacts with air the ball travels through, creating some very interesting effects. The layer stuck to the ball is called the *Boundary Layer*. Wings have boundary layers, by the way. Any object moving through the air can generate one.

Spinning objects are particularly interesting because the boundary layer can be moving in another direction than the center of gravity. Remember, that's how we plot the course of an object; tracking the CG. The point is that The Tube is making lift from spinning, thanks to a boundary layer effect. The reason it curves has to do with *Precession.*

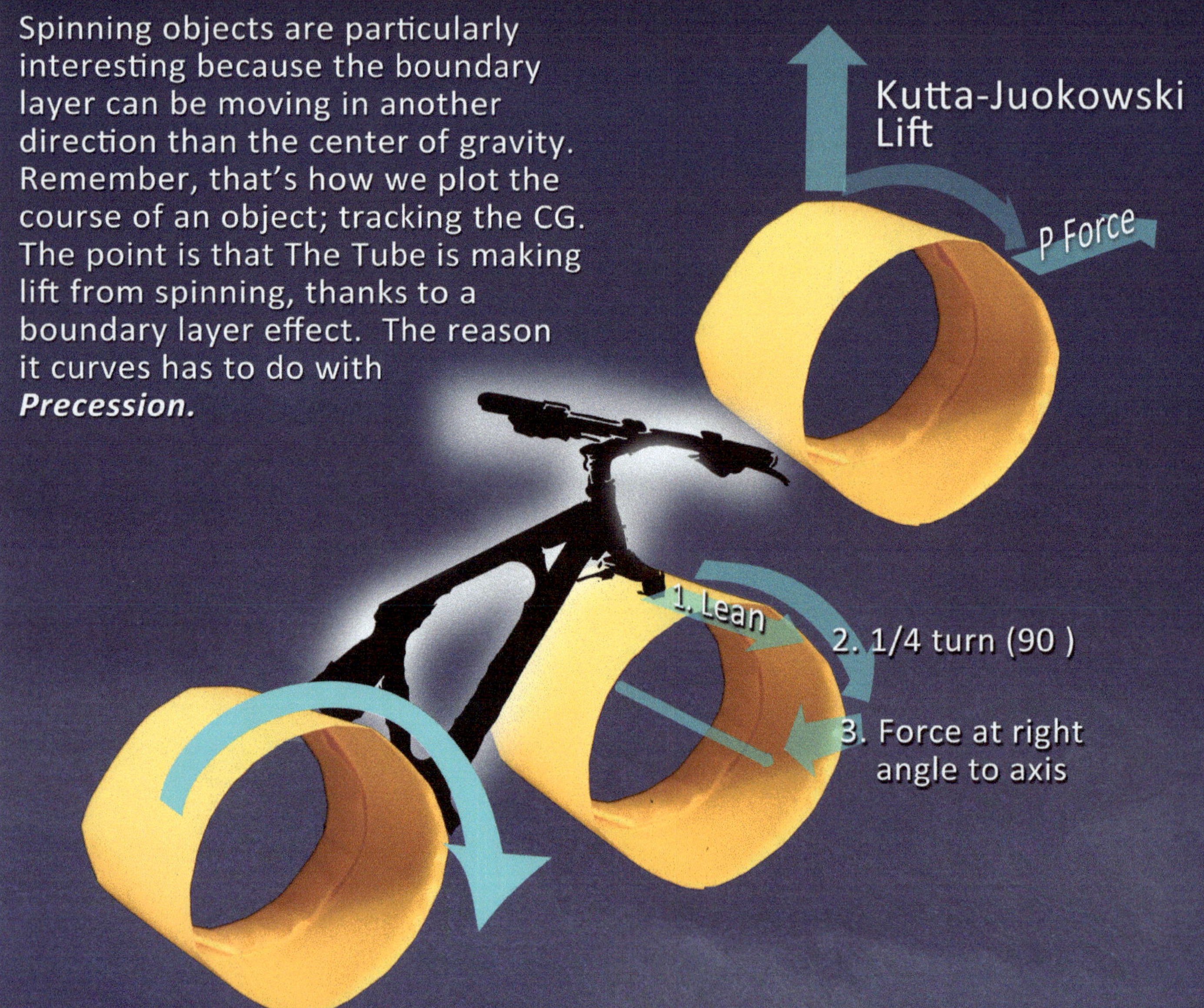

Our tubular bike: Just like a regular bike, to turn right, you lean to the right. One quarter turn later, the force acts at a right angle to the axle, and turns the front wheel to the right. That's precession or P Force.   As the tube flies, the lift forces pulls upward, like the top picture. 1/4 turn later, there's a little pull to the right. The rotational axis on a bike wheel is where the axle sits.

Flap Happy

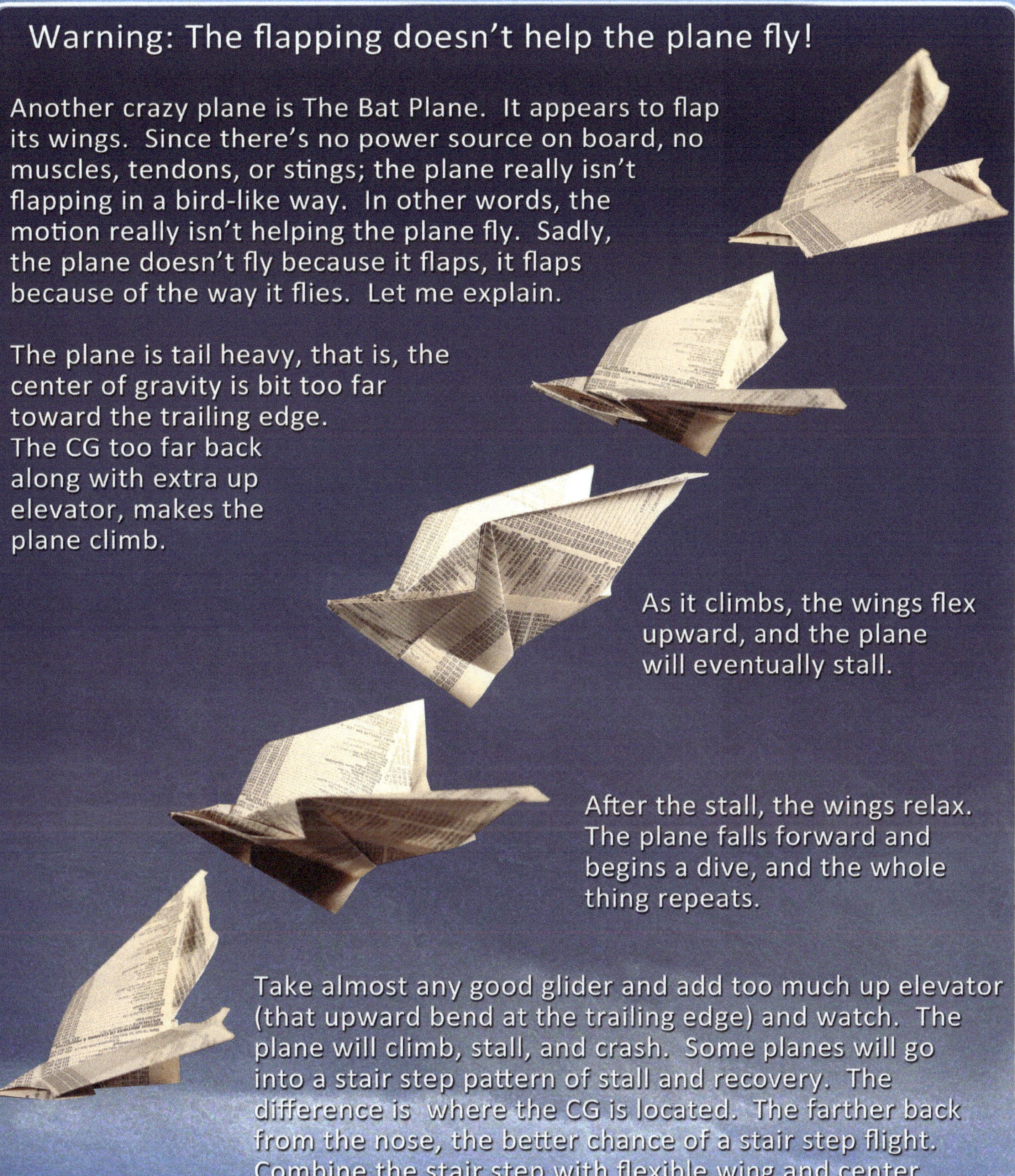

Warning: The flapping doesn't help the plane fly!

Another crazy plane is The Bat Plane. It appears to flap its wings. Since there's no power source on board, no muscles, tendons, or stings; the plane really isn't flapping in a bird-like way. In other words, the motion really isn't helping the plane fly. Sadly, the plane doesn't fly because it flaps, it flaps because of the way it flies. Let me explain.

The plane is tail heavy, that is, the center of gravity is bit too far toward the trailing edge. The CG too far back along with extra up elevator, makes the plane climb.

As it climbs, the wings flex upward, and the plane will eventually stall.

After the stall, the wings relax. The plane falls forward and begins a dive, and the whole thing repeats.

Take almost any good glider and add too much up elevator (that upward bend at the trailing edge) and watch. The plane will climb, stall, and crash. Some planes will go into a stair step pattern of stall and recovery. The difference is where the CG is located. The farther back from the nose, the better chance of a stair step flight. Combine the stair step with flexible wing and center creases, and presto-- a flapping motion.

Fold a Boomerang I from 11 x 17 inch paper. Turn the winglets down and add up elevator where you see it here. Flex the body of the plane a few times. Fine tuning the dihedral and up elevator make it flap better.

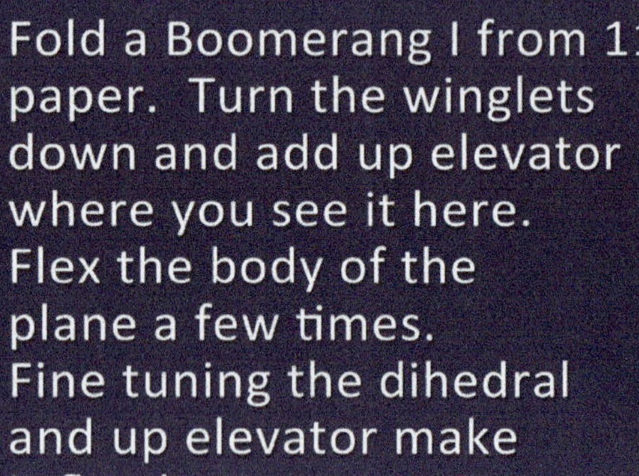

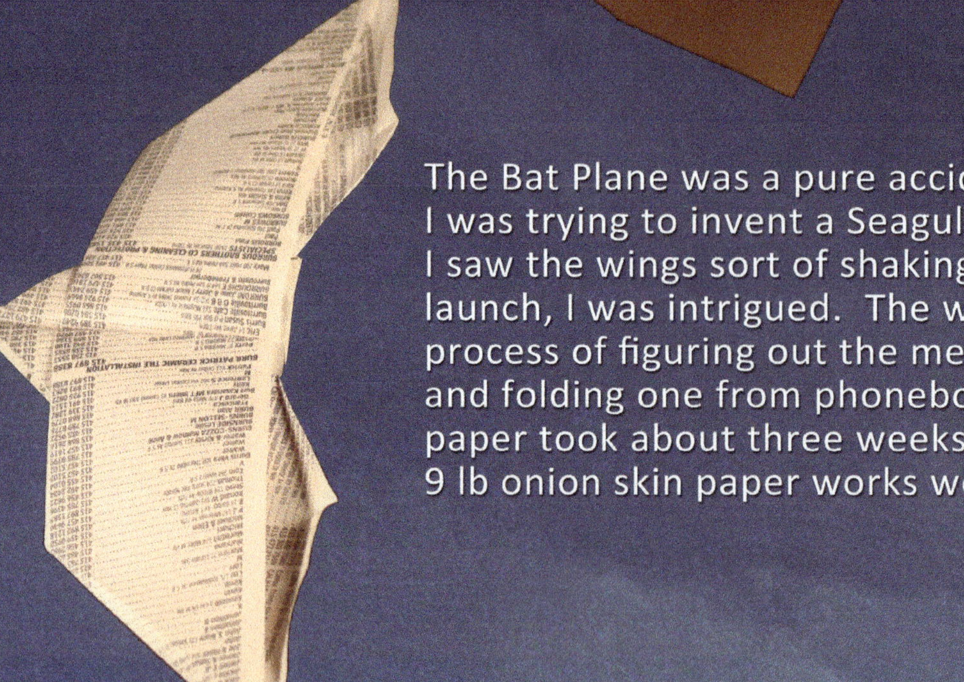

The Bat Plane was a pure accident. I was trying to invent a Seagull. Once I saw the wings sort of shaking on launch, I was intrigued. The whole process of figuring out the mechanics and folding one from phonebook paper took about three weeks. 9 lb onion skin paper works well too.

Canard Designs

A stall resistant aircraft design

The Super Canard has the performance
to match its superb styling. The
small wing in front makes it a
canard design. Note the upward
bend at the front of
the little wing,
which is also
called a canard.

The upward bend adds angle of
incidence (more tilt into the air
flow). As the plane gets close to
a stall, the little wing stalls first.
The loss of lift makes the nose drop.

After the nose drops, the main
wing is no longer in danger
of stalling. It's a very
delicate balance
between up elelvator
and upward bend
on the canard.

Follow Foils

Follow Foils are planes you walk behind (follow) to assist in flying.

We create an updraft by walking forward with the flat side of the cardboard pointed in the direction of travel. If we lean the top of the cardboard back a little bit, we force air to go up the face of the cardboard.

By positioning the updraft correctly under the plane, we can keep it aloft.

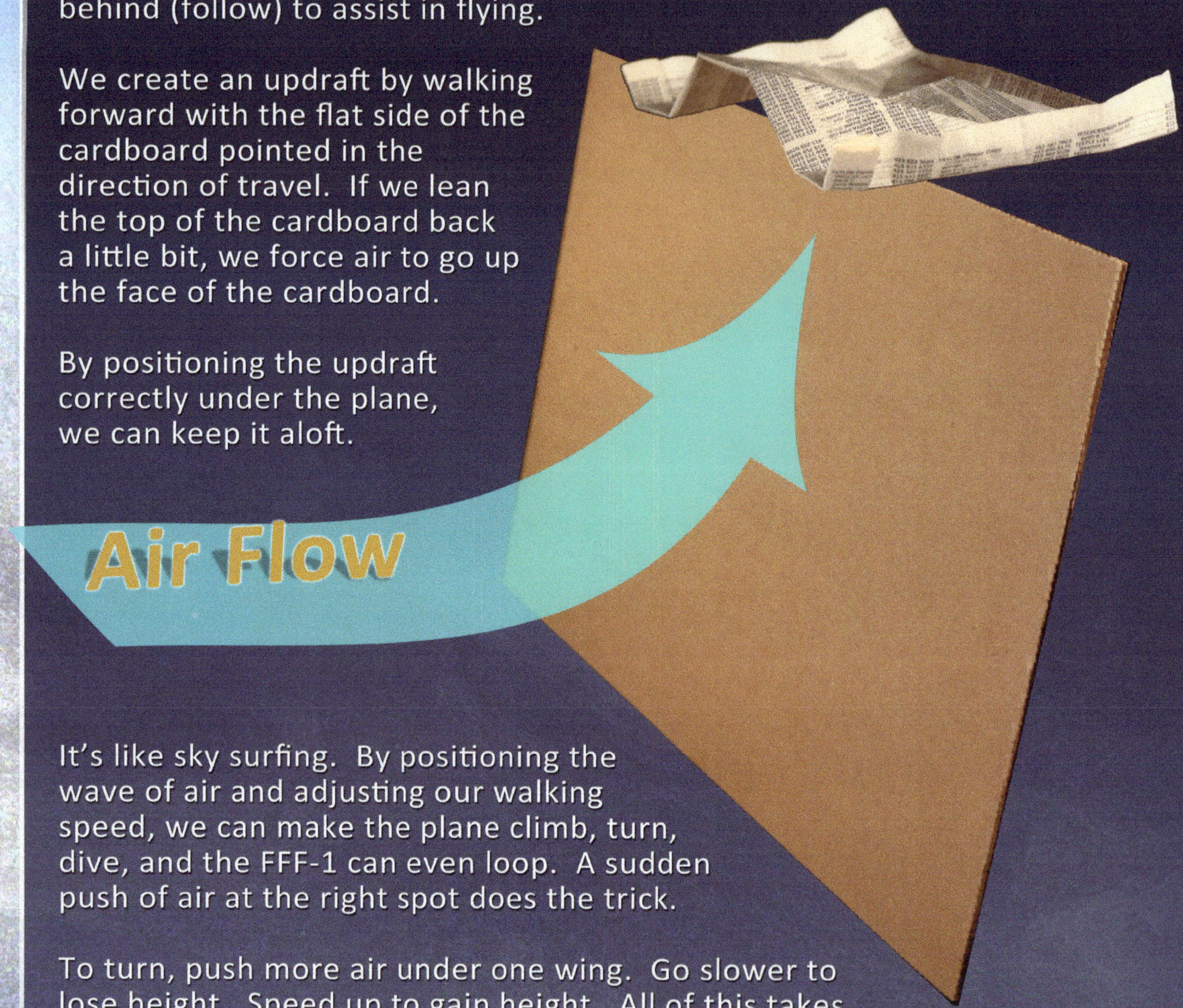

It's like sky surfing. By positioning the wave of air and adjusting our walking speed, we can make the plane climb, turn, dive, and the FFF-1 can even loop. A sudden push of air at the right spot does the trick.

To turn, push more air under one wing. Go slower to lose height. Speed up to gain height. All of this takes practice... lots of practice.

Two Key Factors:

Light weight and low Wing Loading

Follow Foils need to be light weight, or more accurately, have low *wing loading.* Wing loading is simply a measure of how much weight each square unit of wing is lifting.

There's a subtle difference between light weight and wing loading. If I make two planes, each from a sheet of thin graphing paper, each will weigh the same. Each will have a light weight.

The smaller wing will have to lift the same amount of weight as the larger wing by using less lifting surface. Lift is made by airflow around the wing, so the small wing will have to increase airflow to get the same lift. What we're really talking about, is the difference between a dart and a glider. The smaller the wing, the more weight each tiny square of wing will need to lift, and the faster the plane will need to fly. That's the long way of saying, Follow Foils need to be light weight and have low wing loading.

When you use graph paper to make planes, it's easy to actually calculate the wing loading. Count the number of squares on the wing. Divide the weight of the paper by the number of squares, and you get the exact wing loading of the plane. A nice little science experiment might be charting wing size, airspeed, and wing loading.

Sink Rate

First we need low wing loading. Next we need a low *sink rate.*

Sink rate is literally the rate the plane is sinking; moving downward as it glides. All gliders have a sink rate. Slower sink rates are better for Follow Foils. Note that sink rate is not the same as glide ratio.

Glide ratio is a measure of how far forward a plane travels for each foot of height lost. As we'll see, it's possible to have a terrible glide ratio and a great sink rate.

One of the best planes to practice with is The Tumbling Wing. The name is a bit misleading. It doesn't fly like a regular wing. In fact, it just falls in an organized way. Since it's falling, it doesn't really have a glide ratio. It has a downward path of roughly one to one: for every foot it moves forward, it loses a foot of height. That would be a terrible glide ratio for any airplane. But the sink rate is so low, it's one of the easiest planes to fly with the cardboard. Just get it falling in a straight line and practice making it fly farther and farther.

There was actually one aircraft with a one to one glide ratio. It retired a few years back. NASA called the space shuttle the flying brick. That one to one glide ratio was the reason.

BASIC FOLDS

Reverse Fold

Make a crease to mark where the fold will go, and then unfold. Spread the layers wide, and then pull the point down between the layers. Remake the first crease.

Squash Fold

Lift up the flap. Spread the layers open. Press the creased edge down and keep opening the layers wide. Press the crease to the center and flatten.

Waterbomb Base

Make diagonal folds and flip it over. Fold across the X and unfold. Flip over. Press down where all the creases meet. Bring the top down and the sides in.

BASIC FOLDS

Sink Fold

Fold the top down. Unfold everything and flip it over. Follow the little square with mountain folds. Push the little X down. Use the creases to flatten it. Flip it over.

Petal Fold

Fold the lower edges to the center and unfold. Lift the top layer. Keep lifting and form a crease that connects the two new creases. Start reversing the creases on the top layer. Make new creases that come together at the bottom center.

Dart
Stunt
Glider

Easy
Advanced
Skill Level

Path

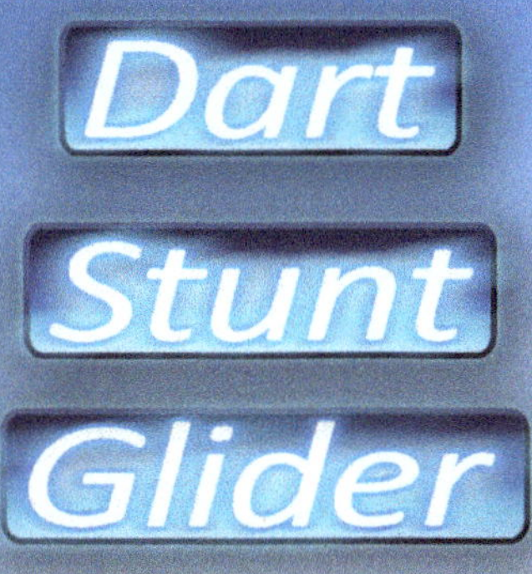

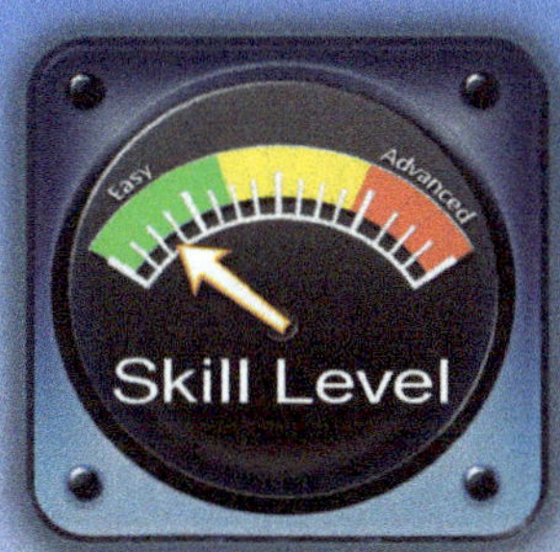

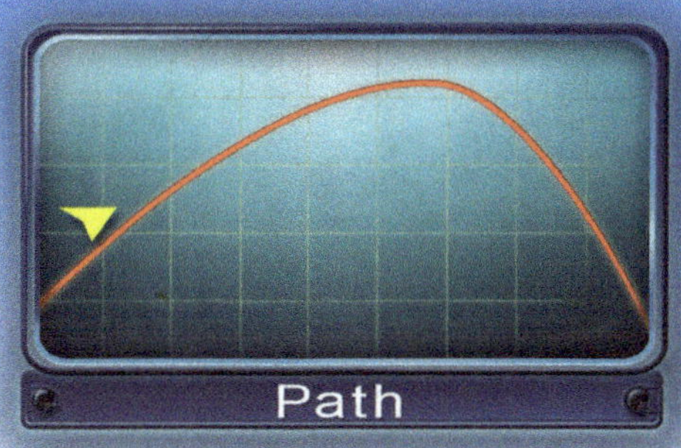

These gauges let you know what kind of plane, how easy it is to fold, and a typical flight path.

Folding Symbols

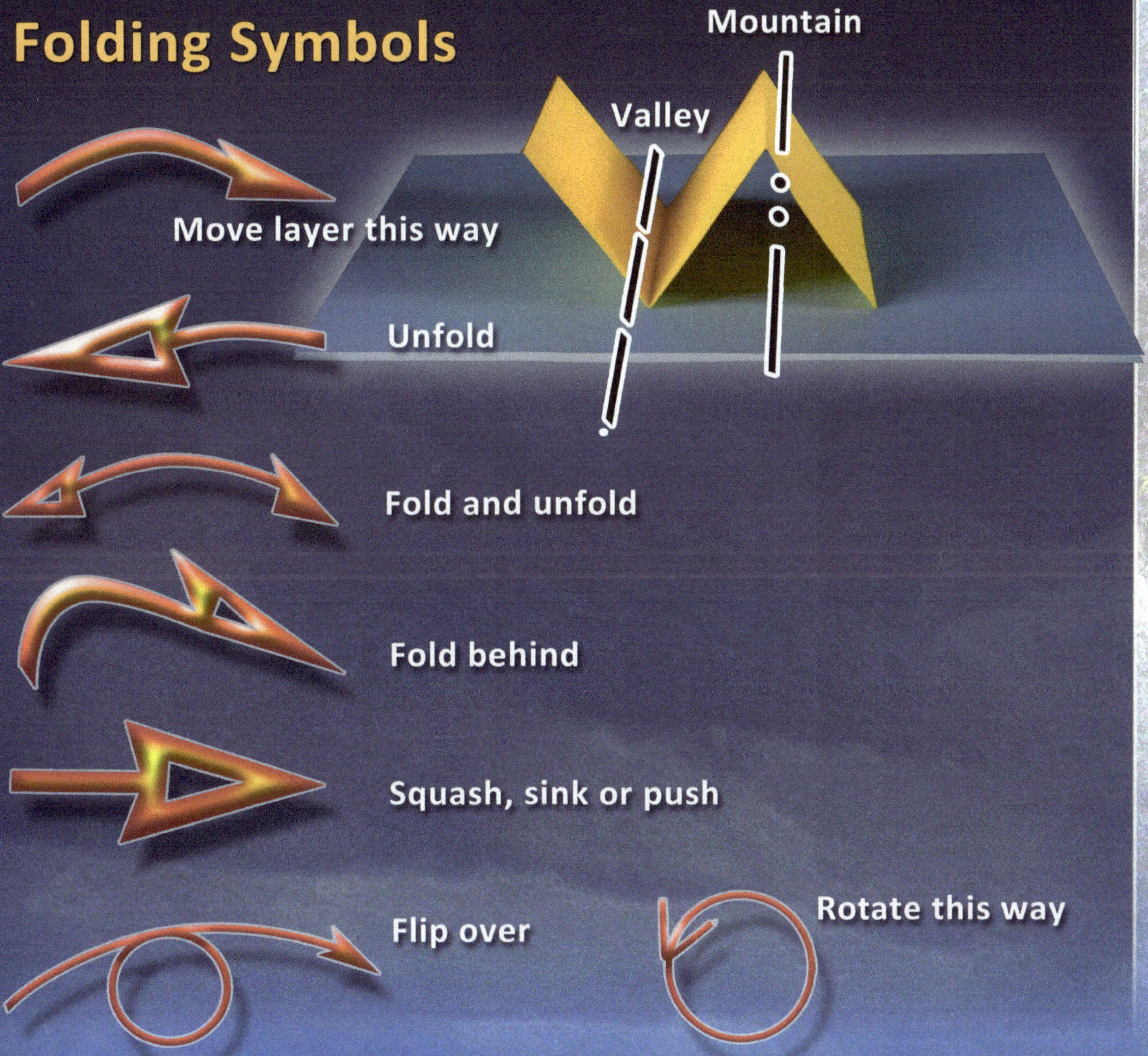

Mountain
Valley
Move layer this way
Unfold
Fold and unfold
Fold behind
Squash, sink or push
Flip over
Rotate this way

Yes, it's in the book!

Every plane from John's
world famous paper airplane program

- The World Record Distance Plane
- All 24 of John's Paper Airplane Program Planes
- Adjusting and Throwing Tricks
- Flight Theory

John Collins and Joe Ayoob smashed the
distance record in 2012, and since then
the world can't get enough of John's paper
airplanes. Now you can fold the amazing collection of paper airplanes
yourself. From the biggest names in Silicon Valley to the smallest classroom
you can imagine, John has thrown these planes for people around the world
Finally, it's your turn!

Contact John through www.ThePaperAirplaneGuy.com

U.S. $17.99
CRAFTS & HOBBIES - PAPERCRAFTS

$17.99
ISBN 978-1-970977-18-9
51799

9 781970 977189